Dictionary of Astronomy, Space, and Atmospheric Phenomena

Dictionary of Astronomy, Space, and Atmospheric Phenomena

David F. Tver

with the assistance of
Lloyd Motz
and
William K. Hartmann

VNR VAN NOSTRAND REINHOLD COMPANY
NEW YORK   CINCINNATI   ATLANTA   DALLAS   SAN FRANCISCO
LONDON   TORONTO   MELBOURNE

Van Nostrand Reinhold Company Regional Offices:
New York   Cincinnati   Atlanta   Dallas   San Francisco

Van Nostrand Reinhold Company International Offices:
London   Toronto   Melbourne

Library of Congress Catalog Card Number: 79-15372
ISBN: 0-442-24045-7

Manufactured in the United States of America

Published by Van Nostrand Reinhold Company
135 West 50th Street, New York, N.Y. 10020

Published simultaneously in Canada by Van Nostrand Reinhold Ltd.

15 14 13 12 11 10 9 8 7 6 5 4 3 2 1

Library of Congress Cataloging in Publication Data

Tver, David F
  Dictionary of astronomy, space, and atmospheric
phenomena.

    1. Astronomy—Dictionaries.   2. Space sciences—
Dictionaries.   3. Atmosphere—Dictionaries.   I. Title.
QB14.T83   520'.3   79-15372
ISBN 0-442-24045-7

Preface

QB14
'T83

From the earliest times, astronomy has been a vital part of science. The knowledge gained by astronomers through the centuries has taught us to appreciate the earth's place in the family of planets that evolve around the sun; it has also made us aware of the insignificance of the sun itself to the known universe.

The frontier today and for time eternal will be outer space. Its magnitude is still beyond comprehension. We are just beginning to take pioneering steps in a continuing expanding adventure that will affect every person on this planet.

So rapid and varied have been the developments in space in the last 50 years, that an entire book can be devoted to each individual aspect of it. Many of these books become obsolete almost before they are printed. Ground-based observing instruments continue to grow in size and sophistication, ranging from the traditional telescope design to the new multiple-mirror telescope (MMT). Automated observations from spacecraft recording information far beyond our atmospheric curtain have opened new windows and increased our knowledge.

New classes of objects beyond our solar system are becoming more known to the general public. We read and hear about such objects as quasars, pulsars, X-ray stars, black holes, proto stars, cosmic bursters, and radio galaxies. We are constantly entering new fields of research, such as neutrino and radio astronomy, and infrared astronomy.

Never before has science moved so fast in gathering information, changing concepts, exploding theories, and developing new hypotheses. New horizons are constantly being opened, and the future in space exploration is almost unlimited.

For these reasons, there is a need by the general public, the student, and amateur astronomer for a dictionary of astronomy. The *Dictionary of Astronomy*, which is designed to aid in the understanding of the general terms that are second nature to professional astronomers, is one of the most comprehensive of its kind. The definitions are simple and concise, but sufficiently detailed where necessary. It should be a valuable reference guide to help everyone interested in astronomy keep abreast of what is happening in the heavens above and the research below.

DAVID F. TVER

A

aberration (or aberration of light). The apparent angular displacement of the position of a star (or other light source) in the direction of motion of the observer, caused by the combination of the velocity of the observer and the velocity of light. The maximum amount of displacement, called the constant of aberration, is about 20″.5 for stars. The amount of this displacement depends on three factors: the speed of the observer relative to the star, the speed of light, and the star's direction relative to the Earth's motion. The displacement is greatest when the Earth moves at right angles to the star's direction and becomes zero if the Earth moves toward or away from the star. If the Earth had only uniform motion in a straight line, the displacement of the star would always be the same and would therefore be unnoticed. But as the Earth revolves around the sun, the changing direction of its motion causes the star's displacement to change direction as well so that the star seems to describe a small orbit.

aberration, chromatic. Imperfect correction of a compound lens, producing a colored fringe around the image of a star.

aberration orbits of the stars. The apparent annual movement of stars due to aberration. A star at the ecliptic pole appears to have a nearly circular apparent orbit because of aberration caused by the Earth's motion in its near circular orbit. A star in the ecliptic oscillates in a straight line. Between the ecliptic and its poles the aberration orbit is an ellipse. The true position of the star lies near the center of these apparent orbits, which do not reflect any actual stellar motion. The numbers in the diagram mark corresponding positions of the Earth in its orbit and of the stars in their apparent aberration orbits. The outer figures show the observed forms of the aberration orbits.

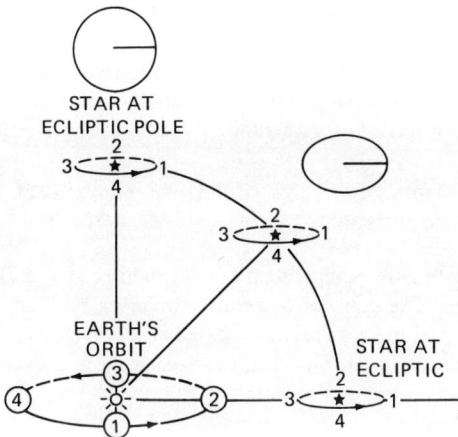

1

aberration, spherical.   A lens imperfection in which the focal plane for rays passing near the center of the lens differs from that for rays close to the edge.

aberration of light.   See aberration.

absolute luminosity.   A measure of the energy generation rate of a star in absolute physical units, such as ergs per second.

absolute magnitude.   A measure of the intrinsic brightness of a star given by the apparent magnitude it would have if it were moved to a distance of 10 parsecs from the observer.

absolute temperature.   A temperature scale in which zero is absolute zero, 273° is water's freezing point, and 373° is water's boiling point. Usually designated as °K or "degrees Kelvin," it is used for most scientific measurements

absolute zero.   A temperature of 0° K at which all thermal motion of atoms and molecules theoretically ceases, according to the kinetic theory of ideal gasses. It equals −459.7° F, and −273.15° C.

absorption band.   A range of wavelengths (colors or frequencies) in the electro-magnetic spectrum within which radiant energy is absorbed as light passes through substances. Different substances display different bands. Absorption bands generally are caused by molecules, whereas absorption lines are caused by atoms.

absorption line.   A minute range of wavelengths (color or frequency) in the electro-magnetic spectrum within which radiant energy is absorbed by a substance through which it is passing. Each line is associated with energy loss from the light beam associated with changes in electronic structures of atoms as they are struck by photons from the beam.

absorption line spectrum.   A spectrum interrupted by dark lines. Absorption spectra are generally created when cooler gas intervenes between the hot source of a continuous spectrum and the observer. A certain pattern of dark lines is unique to each gaseous element. Sunlight, having passed through the atmosphere of both the Sun and the Earth, produces a dark-line spectrum.

abundances of elements.   Measures of relative numbers of atoms of different elements in a specified material.

acceleration.   Speeding up or slowing down; more precisely, a change in amount or direction of velocity. The concept of velocity involves both speed and direction of motions; thus acceleration may appear as changing speed, changing direction, or both. A planet moving in a circular orbit illustrates acceleration in direction only; the speed is constant. Increasing speed of a rocket during vertical launch illustrates acceleration by change of speed only.

acceleration due to gravity.   Acceleration of one mass toward another due to gravitational attraction. Usually indicates the acceleration at the surface of a specified planet. On Earth it is approximately 980 cm/sec^2.

accelerator (telescope).   A compound that speeds up the setting of a polyester resin, usually cobalt naphthenate.

accretion.   Any mechanism, such as gravity or mechanical sticking, by which two colliding objects remain bonded to each other, or by which net mass gain is the result of collision.

achondrite.   A type of meteorite without chondrules, more altered from the primitive meteorite state than chondrite meteorites.

achromatic lens.   A lens which produces optical images free from false color. These are generally made by combining separate lenses, called elements, made from different kinds of glass. A single element lens focuses different wavelengths at different distances so that a clear image of a star is not obtained anywhere. A second element can be designed to improve the image.

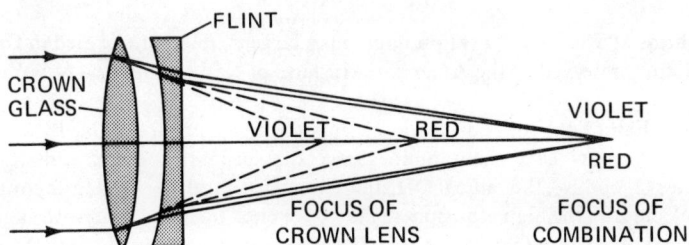

Achromatic telescope.   Refractory telescope with an achromatic objective lens. The design was introduced in 1758 and led to greatly improved astronomical observation.

aclinic line.   The line through those points on the Earth's surface at which magnetic dip is zero. The aclinic line is a particular case of an isoclinic line.

active sun.   The sun during portions of the sunspot cycle when the numbers of sunspots and flare activities are high.

adiabatic.   Adjective describing a process in which no energy is added or removed from a system by an external source.

Adonis.   A small asteroid with a diameter of a few miles. Owing to the eccentricity of its orbit, it came within 1½ million miles of the Earth in 1936.

advection.   The process of the transport of an atmospheric property, such as heat, solely by horizontal mass motion of the atmosphere; also, the rate of change of the value of the advected property at a given point.

aerolite.   A stony meteorite (term no longer commonly used).

aeronomy.   The science of the atmosphere of any celestial body, especially in relation to its reaction to radiation bombardment from outer space. The study of the upper regions of the atmosphere where ionizations, dissociations, and chemical reactions take place.

aerospace.   The Earth's envelope of air and the space above it in which space vehicles fly.

afterglow.   A broad, high arc of radiance or glow seen occasionally in the western sky above the highest clouds in deepening twilight, caused by the scattering effect of very fine particles of dust suspended in the upper atmosphere. Often pronounced in months after major volcanic eruptions inject dust into the high atmosphere.

agonic line.   A line joining points at which the magnetic variation is zero. The agonic line is a particular case of an isogonic line.

Ahmighito.   "The Tent." Eskimo name for a large meteorite recovered in Greenland in 1897 and preserved in the American Museum of Natural History, New York City.

airglow.   Faint light emission from the upper atmosphere detectable by instruments at night. Airglow is a chemiluminescence due primarily to the emission of the molecules O_2 and N_2, the radical OH, and the atoms O and Na. During daytime, atoms and molecules in the high atmosphere may be excited by interaction with sunlight. At night when the exciting source is absent the atoms and molecules revert to their ground state with different time scales, producing much of the airglow. A similar glow was found in Venus in 1976 by Russian spacecraft. On earth the glow appears at altitudes of about 100 to 200 km. When first detected it was called the "permanent aurora." It gives us twice as much light as do all the stars combined. It places a limit on the duration of exposure in direct celestial photography before the plates become hopelessly fogged. The glow is faintest overhead and reaches its greatest intensity 10° above the horizon where we look through a greater thickness of air.

air mass.   A measurement of the relative amount of air that a light beam traverses before reaching the observer, usually expressed as a multiple of the length of the path at the zenith. It rapidly increases toward the horizon and causes stars to dim as they approach the horizon. Photometric observations of stars in different parts of the sky must be corrected for this effect.

air shower.   A grouping of cosmic-ray particles observed in the atmosphere; a cascade shower in the atmosphere. Primary cosmic rays slowed down in the atmosphere emit bremsstrahlung photons of high energy. Each of these photons produces secondary electrons which generate more photons, and the process continues until the available energy is absorbed.

airy disk. In a telescope, the false or spurious disk which forms the image of a star. It has nothing to do with the star's actual disk which is too small to be seen, but is caused by diffraction of light. It is named after its discoverer, George Airy.

Aitken nuclei. Microscopic, atmospheric particles, usually of dust, which serve as condensation nuclei for droplet growth during condensation. These nuclei are both solid and liquid particles whose diameters are of the order of tenths of microns or even smaller. First studied by John Aitken.

al. Arabic article incorporated as the first syllable of many technical words transmitted to European culture by the Arabs during the Middle Ages.

Al-Battani, Muhammad. (ca. A.D. 900) Arab astronomer known in European literature as Albategnius.

albedo. A measure of the light-reflection power of any surface but usually that of a planet, satellite, or asteroid, compared with an ideal white matte surface which absorbs no light, the albedo of which is unity. Typical albedoes are: Earth, 0.4; Moon, 0.11; Mars, 0.15; Venus, 0.59; asteroids, 0.1. Since natural surfaces reflect different amounts of light in different directions, specific types of albedoes specify whether measurement is made in one direction or averaged over all directions. Above values are technically called geometric albedoes (q.v.).

Alberio. See Beta Cygni.

Alcor. Faint companion of Mizar in Ursa Major, it is a well-known visual double. Magnitude 3.9.

Aldebaran. See Alpha Tauri.

Alfven speed. The speed at which Alfvén waves are propagated along the magnetic field. For a perfectly conducting fluid with a mass density of one kilogram per cubic meter in a magnetic field of 10,000 gauss, the Alfven speed is about 1,000 meters per second, while the speed of sound in air is about 300 meters per second.

Alfvén wave. A transverse wave in a magnetohydrodynamic field in which the driving force is introduced by the magnetic field.

Algol (B Persei or the Winking Demon Star). An eclipsing binar and one of the best observed stars in the sky. It has an orbit period of 2.867 days and the primary minimum lasts for 14 percent of the period, or 9.9 hours. The orbit is practically circular and inclined 81° to the plane of the sky. During most of its period Algol looks like a 2nd-magnitude star, but every 2 days, 13 hours, it starts to fade, dimming steadily for 3 1/2 hours until it is reduced to below magnitude 3. It remains in this state for 20 minutes after which it starts to brighten once more, taking another 3 1/2 hours to regain its normal brilliance. Algol was considered by some ancients as a most unfortunate and dangerous star in the heavens, always associated with violence and sudden death. Algol is about 88 light-years distant, and each of the two co-orbiting stars is about 3 1/2 times larger than the Sun.

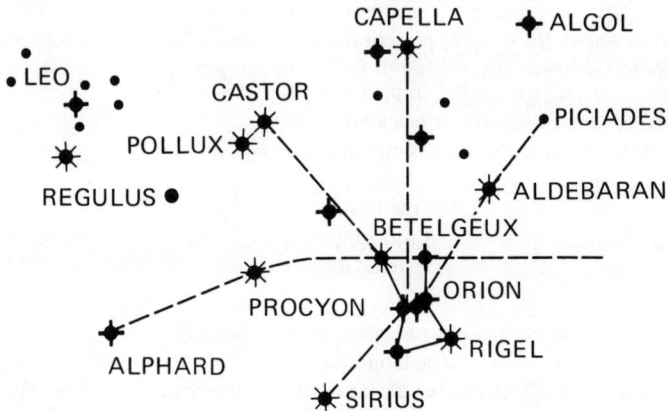

algol variable.   An eclipsing variable in which two components are sufficiently separated for them to be essentially spherical in shape. Named after the prototype β Persei. The light curves are characterized by having comparatively flat maximum.

alidade.   A rule equipped with a visual or telescopic sight, used to determine direction.

Alioth (Ursa Major).   The third star from the end of the handle is Alioth, the name meaning "the fat tail of the eastern sheep." It is a double, the two stars revolving around a common center of mass in a little over four years. It is the brightest star in the constellation and is a visual magnitude 1.8.

Almagest.   One of the early star catalogs, usually referring to the Alexandrian astronomer Claudius Ptolemy (ca. A.D. 140). Its name, which signifies The Greatest Work, comes from the Arabs who preserved it for posterity.

almanac.   Any guide to events during a specified year.

Alpha Aquilae.   Named Altair from part of the Arabic name of the constellation, and means "the flying vulture." Prominent in summer evening skies, it has a visual magnitude 0.77. It is a main-sequence A star. Estimated distance is about 16 light-years.

Alpha Arae. Called by Arabic name Hamal, meaning "the sheep." A star of visual magnitude 2.0, estimated distance 75 light-years. It is a K giant.

Alpha Boötes. Arcturus, a prominent reddish K giant of visual magnitude –0.1. It is prominent in the spring and summer evening skies, and is frequently mentioned in ancient literature. It is estimated to be about 36 light-years away. It has a high proper motion of 2.3 seconds per year, and has moved more than one lunar diameter in apparent position since the days of Ptolemy.

Alpha Canis Majoris. The Dog Star, Sirius, the brightest star in the sky in terms of apparent brightness. It is a hot, bluish-white star of spectral class A, on the main sequence of the H-R diagram. It has a visual magnitude of about –1.45 and is about 8.8 light-years away. Orbiting around it every 50 years is a white dwarf with the approximate mass of the Sun but only 3 percent of the Sun's radius. The white dwarf was predicted in 1844 by the German astronomer Friedrich Bessel, who deduced its presence from Sirius's motions; it was first seen in 1862 by American telescope maker Alvin Clark. It was identified as a dense dwarf (a type hitherto unknown) in 1915 by Mt. Wilson observer W. S. Adams.

Alpha Canis Minoris. The Little Dog Star, Procyon, a rival to the Dog Star, Sirius, which is nearby in the sky. Like Sirius, it has a white dwarf companion orbiting around it. It has a visual magnitude of about 0.4 and is estimated to be about 11 light-years away.

Alpha Capricorni. A naked-eye double of magnitudes 3.8 and 4.6. Named Giedi, each of the two stars in turn has a companion of about 9th magnitude. Giedi 2 is a triple system.

Alpha Cassiopeiae. Called Schedar by the Arabs, meaning "the Beast." It is a K giant of visual magnitude 2.2, reddish hue, about 150 light-years distant. It has a telescopic companion of the 9th magnitude.

Alpha Centauri. The third brightest among the visible stars, and probably the second nearest among our star neighbors. It was one of the first whose distance, 4.3 light-years, was determined. It is a double star, the components being magnitude –0.1 and +1.4. The two together are about double the mass of the Sun, the brighter component being somewhat the larger. The brighter component matches the Sun very closely in actual brightness, mass, and color, being a G2 main-sequence star. They revolve around a common center of mass in highly eccentric orbits in a period of 80 years. This star was for many years an object of worship along the River Nile. It is only rarely designated by its Arab name, Rigel Kent.

Alpha Corona Borealis. Known as Alphecca, a variable star of average magnitude 2.2.

Alpha Cygni.   The star at the head of the Northern Cross called Deneb by the Arabs, meaning "the Hen's Tail." It is a brilliant white A supergiant of visual magnitude 1.25 lying in the densest part of the Milky Way, its brilliance contrasting with the soft glow of the millions of faint stars lying far beyond it. With Deneb and Vega it forms part of the "summer right triangle" of prominent first magnitude stars, the latter occupying the apex. Deneb is estimated to be 1600 light-years distant. The spectroscope shows it to be accompanied by a faint star of the 12th magnitude.

alpha decay.   The radioactive transformation of an atomic nucleus by an alpha particle emission. The product is a nucleus having a mass number four units smaller, and an atomic number two units smaller, than the original nucleus.

Alpha Leonis.   The bright star, Regulus, known as the heart of Leo, the Lion. Regulus has visual magnitude 1.3. It is a B-type hot main-sequence star of bluish color, located at a distance of about 85 light-years. It is variable and has two companions, probably optical and not physical, at separations of 4″ and 217″.

Alpha Lyrae.   Vega is a bluish white star with apparent visual magnitude 0.0. It has spectral type A and is a hot star more massive than the Sun and relatively close, with a distance of 26 light-years. Due to precession of the Earth's equinoxes, Vega was the North Pole Star about 14,000 years ago and will be again in roughly 12,000 years. The star is also of interest because our solar system is traveling toward it, relative to the average positions of other nearby stars, at a speed of about 19 km/sec. In roughly 1/2 million years the solar system will reach the general vicinity of Vega, but Vega is unlikely to be closer than some thousands of Astronomical Units from us.

Alpha Ophiuchi.   Called Ras Alhague by the Arabians, meaning "head of the Serpent Charmer." It is of magnitude 2.2 and its distance is estimated to be 59 light-years. It is a bluish star, 18 times more luminous than the Sun.

Alpha Orionis.   One of the best known stars, called Betelgeuse by the Arabs. It lies on the left shoulder of Orion, the hunter. As seen by the observer, Alpha glows with a magnificent orange tinge, and month to month observations show that it undergoes definite variations in brightness. The fluctuations are slow, but since the approximate range is from magnitude 0.1 to 1.4, they become obvious enough after a time. It lies near the bright orange star Aldebaran in Taurus, whose magnitude is 0.78, making it a useful comparison star. Betelgeuse is an M2 supergiant at an estimated distance of 650 light-years.

alpha particle.   A positively charged particle composed of two protons and two neutrons. An alpha particle has an atomic weight of 4 and a positive charge equal in magnitude to 2 electronic charges; it is essentially a helium nucleus.

Alpha Pegasi.   Alpha on the southwest corner of the great square of Pegasus was called Markab by the Arabs in whose language the word means "saddle." It is situated at the junction of the animal's wings and shoulder. A B giant of visual magnitude 2.5, its distance is estimated at 110 light-years.

Alpha Persei. The central star in Perseus. The Arabs called it Mirfak, meaning "the elbow." It is an F supergiant of visual magnitude 1.8, flanked on either side by stars of the third magnitude. It is a double, the two components revolving around each other in a period of four days. It is estimated to be 520 light-years distant.

Alpha Scorpii. Antares, a brilliant reddish M supergiant of visual magnitude 1.0. It is one of the largest known stars, estimated to be roughly 600 million km in diameter, larger than Mar's orbit. It is estimated to be about 420 light-years away. It has a faint whitish companion of about 5th magnitude, visible in a telescope at a distance of 3 seconds of arc.

Alpha Tauri. The principal star of Taurus known by its Arabic name, Aldebaran. It is a markedly red star of the visual magnitude 0.85 and 14th in the list of the brightest stars in the heavens. The Babylonians called it the "Leading Star of Stars" and the Persians highly regarded it as one of the four "Royal Stars." Aldebaran rises about an hour after the Pleiades, and was often called "The Follower." It lies along the Moon's path and occultations with that body occasionally occur. It is a K giant around 45 times bigger and 360 times brighter than the Sun. Its distance is estimated to be 68 light-years.

Alpha Virginis. Known as Spica, it marks the sheaf of wheat which the Virgin holds in her right hand. An easy way to locate this star is to follow the curves of the three stars in the handle of the Big Dipper to Arcturus, then continue the curve for about the same distance to find Spica. Its visual magnitude is 1.0, and it is about 260 light-years from the sun. It is a B star.

alphabet, Greek.

A	α	Alpha	H	η	Eta	N	ν	Nu	T	τ	Tau
B	β	Beta	Θ	θ	Theta	Ξ	ξ	Xi	Y	υ	Upsilon
Γ	γ	Gamma	I	ι	Iota	O	o	Omicron	Φ	ϕ	Phi
Δ	δ	Delta	K	κ	Kappa	Π	π	Pi	X	χ	Chi
E	ϵ	Epsilon	Λ	λ	Lambda	P	ρ	Rho	Ψ	ψ	Psi
Z	ζ	Zeta	M	μ	Mu	Σ	σ	Sigma	Ω	ω	Omega

Alphecca. See Alpha Corona Borealis.

Alphonsine tables. Tables of predicted planetary positions calculated from Ptolemy's theory in 1242–1262 under sponsorship of King Alphonso X of Castile. The tables' inconsistencies between the positions predicted and the positions actually observed in following centuries contributed to the Copernican revolution.

Altair. See Alpha Aquilae.

altazimuth mounting. A mounting consisting of two axis at right angles to each other. One axis is vertical, the other horizontal.

altitude. Angle of elevation of a star above horizon.

altitude (of north celestial pole).   Altitude of north celestial pole is numerically equal to the latitude of the observer in the northern hemisphere.

aluminizing.   The process of coating a mirror with a thin layer of aluminum to provide a reflective coating. The mirror to be aluminized is thoroughly cleaned and enclosed in a chamber which can be evacuated almost completely. The aluminum is evaporated by means of an intense electrical discharge and deposited on the surface of the mirror.

American Epheeris And Nautical Almanac.   A volume compiled by the staff of the U.S. Naval Observatory since 1855. It contains for each year the positions of the Sun, moving planets, and certain of their satellites and other astronomical information and is used in virtually every astronomical observatory.

Amor asteroids.   See Asteroids, Amor.

analemma.   A graphic scale, usually drawn in the form of a figure eight, showing the latitude of the overhead Sun for every day of the year, as well as the difference between apparent solar time and mean solar time.

Andromeda.   It is near the meridian during early November evenings. In the mythology of ancient Greece, Andromeda represents the chained daughter of Cepheus and Cassiopeia, who is rescued from the sea monster Cetus by the hero Perseus. This large northern constellation was recognized by Ptolemy, who assigned 23 stars to it. The chief stars are Alpha or Alpheratz (2.4), Delta (3.2), Beta (2.0) and Gamma (2.1). They present an interesting color contrast: Alpheratz is white, Gamma is slightly orange, and Beta is definitely orange-red. Andromeda is celebrated because it includes the great spiral galaxy M31.

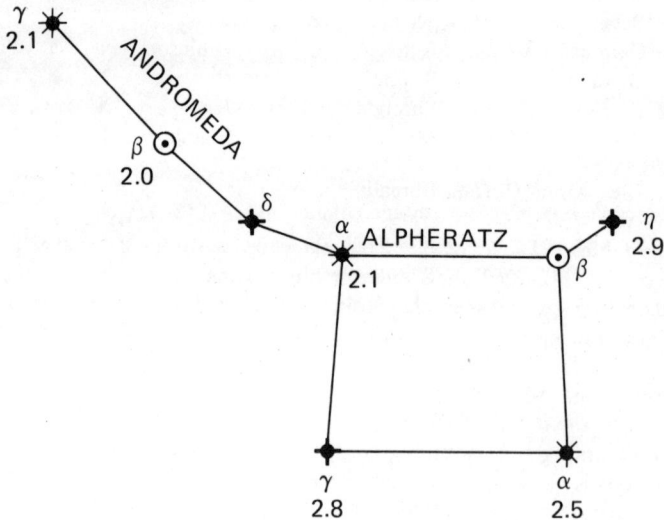

Andromeda galaxy M31.   The brightest and possibly nearest of the exterior spiral galaxies, it can be seen by the naked eye as an elongated hazy glow. Only the central region appears to the naked eye and, for the most part, to the eye at the telescope. Fainter surrounding parts come out clearly on photographs, where this galaxy is shown in its true character as a spiral inclined about 13° to the line of site. Because of its inclination, the nearly circular spiral appears oval to us. It has a distance of 670 kpc and a linear diameter of about 40 kpc. Its spiral pattern and mass seem to resemble closely those of our own spiral galaxy.

Andromedids.   A meteor shower associated with the comet Biela, which occurs yearly about November 14th.

angle.   (1) Angular distance measured in degrees, minutes, and seconds of arc. A full circle is 360°; *1° = 60′* (minutes of arc), *1′ = 60″* (seconds of arc). (2) The inclination to each other of two intersecting lines measured by the arc of a circle intercepted between the two lines forming the angle, the center of the circle being the point of intersection.

angstrom.   The unit in which the wave length of visible light is generally expressed. One angstrom unit is defined as 10^{-8} centimeters. Angstroms are indicated by Å or λ, the Greek letter lambda. Yellow light has a wavelength of about 0.00006 centimeters, which is expressed as 6000Å or λ6000, equal to 0.00024 inch. Named in honor of A. J. Ångstrom, the Swedish physicist who first made accurate measures of the wavelengths of light.

angular diameter.   The apparent diameter of an object measured in degrees or other angular units.

angular distance.   The apparent distance between two objects measured in degrees or other angular units.

angular momentum.   The angular momentum of a moving body, such as a planet revolving around the Sun or a particle of a rotating globe, is the product of the mass, the square of the distance from the center of motion, and the rate of angular motion. The principle of the conservation of angular momentum asserts that the total angular momentum (the sum of these products) of an isolated system is always the same. In the solar system the Sun's rotation is only 2 percent of the total angular momentum, whereas Jupiter's revolution accounts for 60 percent, and the four giant planets carry nearly 98 percent of the total. This fact contradicted many early theories of solar system origin.

angular motion.   The motion of a celestial body measured in units of angle/time, such as a degree/hour, or seconds of arc/year.

annular eclipse.   An eclipse of the Sun in which the Moon obscures the central part of the Sun's disk, leaving a thin ring of light showing round the circumference. An annular eclipse occurs when the moon is directly between us and the Sun but near its farthest point from earth, so that the umbra of the shadow does not reach the Earth.

anomalistic month.   The average period of revolution of the Moon from perigee to perigee, a period of 27 days, 13 hours, 18 minutes, 33.2 seconds.

anomalistic period.   The interval between two successive perigee passages of any satellite in orbit about a primary.

anomalistic year.   The time between two successive passages of the Earth through perihelion. The period of one revolution of the Earth about the Sun from perihelion to perihelion; 365 days, 6 hours, 13 minutes, 53 seconds in 1900 and increasing at the rate of 0.26 seconds per century.

anomaly.   In celestial mechanics, the angle between the radius vector of an orbiting body from its primary (the focus of the orbital ellipse) and the line of apsides of the orbit, measured in the direction of travel, from the point of closest approach to the primary (perifocus).

anorthosite.   A type of igneous rock related to basalt (q.v.) but lighter in color and composed almost entirely of feldspar minerals. It became important in planetary studies when it was discovered by Apollo astronauts to be a major constituent of the lunar uplands.

ansae.   The portions of Saturn's ring system or a distant galaxy extending outside the main disk, as seen from the Earth or any other distant point.

antapex.   The point on the celestial sphere for which the Sun and entire solar system is retreating at about 19.7 km/sec.

Antares.   See Alpha Scorpio.

Antillia (The Air Pump).   An inconspicuous southern constellation near the southern rift in the Milky Way. It contains no star brighter than the 4th magnitude.

antipode.   That point on the celestial sphere or on a planet 180° from another specified point.

antisolar point.   That point on the celestial sphere 180° from the Sun.

antoeci.   Two places in the same meridian and in corresponding latitudes north and south of the Equator.

apastron.   That point of the orbit of one member of a binary star system at which the stars are farthest apart. That point at which they are closest together is called periastron.

aperture.   The diameter of light-collecting lens or mirror in a telescope.

apex of the sun's way.   The point toward which the sun is moving at 19.7 km/sec, relative to nearby stars. It is near Vega, at R.A. $11^h.3$ and Dec. $+30°$.

aphastron.   See apastron.

aphelion.   The point farthest from the Sun in an orbit in the solar system.

apogee.   The point farthest from the Earth on the elliptical orbit of a satellite. The point on the orbit of the Moon or an artificial earth satellite which is farthest from the earth.

Apollo.   (1) A small asteroid which in 1932 came within 12 million km of the Earth; it was lost as it receded again. (2) The United States' manned-lunar landing program. Six successful landings were made from June 20, 1969 (Apollo 11) to Dec. 11, 1972 (Apollo 17).

Apollo asteroids.   See Asteroids, Apollo.

apolune.   The point on the elliptical orbit of a Moon's satellite farthest from the Moon.

apparent diameter.   See angular diameter.

apparent force.   A force introduced in a relative coordinate system in order that Newton laws be satisfied in this system. This force must be equal and opposite to an ecceleration in an inertial coordinate system, in which Newton laws are (by definition) satisfied. Examples are the corioles force and the centrifugal force incorporated in gravity.

apparent motion.   Motion relative to a specified or implied reference point which may itself be in motion. Also called relative motion.

apparent noon.   Moment when the sun crosses the meridian, not necessarily consistent with noon in mean solar time.

apparent position.   The position or the celestial sphere at which a heavenly body can be seen by an observer from the center of the earth at a particular time.

apparent solar day.   The duration of one rotation of the earth on its axis with respect to the apparent Sun. It is measured by successive transits of the apparent Sun over the lower branch of a meridian. The length of the apparent solar day is 24 hours of apparent time and averages the length of the mean solar day, but varies somewhat from day to day.

apparent solar time (AST).   Time measured by the position of the Sun at any point.

apparent sun.   The Sun as it appears in the sky; also called "true sun" and distinguished from the "mean sun"—the position which the Sun would occupy if it appeared to move at a constant angular rate.

apparent time.   Apparent solar time.

Appleton layer.   The F-layer of the earth's ionsphere.

appulse.   (1) The near approach of one celestial body to another on the celestial sphere (i.e., as in occulation, conjunction). (2) A penumbra eclipse of the moon.

apsidal rotation.   A slow rotation of the line of apsides in a binary system due to perturbative gravitational forces.

Apside (line of).   The longest axis of an elliptical orbit, and therefore the line which passes through the two foci.

apsis.   In celestial mechanics, either of the two orbital points on the line of apsides. See preceding entry.

Apus (The Bird of Paradise).   A faint southern constellation distinguished by a small triangle of 4th magnitude stars near the southern celestial pole. The constellation is usually attributed to Johann Bayer in 1603, but he may have adopted it from the accounts of voyagers to the southern hemisphere in the previous century who would be acquainted with the bird of paradise it is supposed to represent.

Aquarius (The Water Pourer).   Zodiacal constellation in the evening sky in Autumn. Well-known in astrological superstition, it is irregularly shaped and lacking prominent asterisms. Classically, this star group was supposed to outline the figure of a man pouring water from a jar or urn.

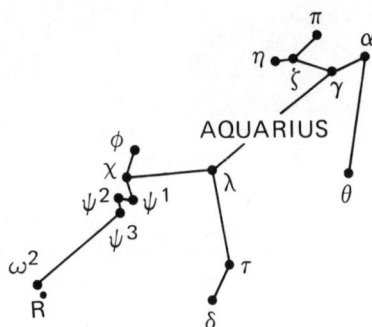

Aquarius, Age of.   Era during which vernal equinox, which moves due to precession, lies in the constellation or astrological sign of Aquarius. Because of precession and changes in the definition of the constellation boundary, the start of the Age of Aquarius is defined imprecisely by many decades.

Aquarius (Astrology).   According to astrological superstition those born under the sign of Aquarius were said to be progressive, humanitarian, interested in education and strongly inclined toward the occult. The two wavy lines used as its symbol represented not only the undulating movements of waves but, to some astrologers, two serpents traveling in opposite directions, illustrating the true balance between reason and intuition, by which wisdom is acquired. It was taught that reason alone leads to materialism, while intuition alone tends to mysticism. The two, properly combined, lead to understanding.

Aquila (The Eagle).   A constellation said to represent an Eagle in full flight. It can be recognized easily by its row of three bright stars equally spaced, the brightest of which is 1st magnitude Altair, in the center.

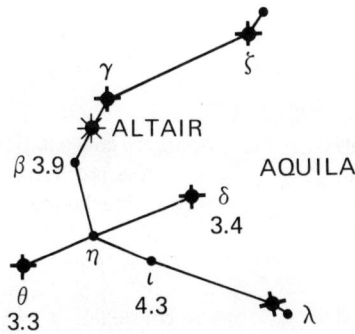

Ara (The Alter).   South of the Scorpion's tail is the small constellation of Ara. One of the ancient star groups listed by Ptolemy, it is represented as an altar, the smoke from which may be imagined in the streaming starlight of the Milky Way. It was also the altar of sacrifice. Chief stars include Alpha (magnitude 2.9), Beta (2.9), and Gamma (3.3).

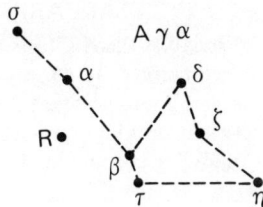

Arabic Alphabet.

'	ا	Alif	glottal catch.
b	ب	Bā	
t	ت	Tā	
th	ث	Thā	
j	ج	Jīm	like *j* in *Jack*, or *g* in *gem*.
ḥ	ح	Ḥā	smooth guttural aspirate.
ḫ	خ	Ḫā	like *ch* in the Scotch word *loch*; in the German *rache*. Velar spirant.
d	د	Dāl	
dh	ذ	Dhāl	like *th* in *the, that*.
r	ر	Rā	
z	ز	Zāy	
s	س	Sin	
sh	ش	Shīn	
ṣ	ص	Ṣād	like *ts*; or, as in modern Arabic, a sharp palatal *s*.
ḍ	ض	Ḍād	*d* with a glottal catch.
ṭ	ط	Ṭā	emphatic palatal *t*.
th	ظ	Ṭhā	emphatic *z*.
'	ع	'Ain	strong glottal catch.
gh	غ	Ghain	post-palatal guttural.
f	ف	Fā	
ḳ	ق	Ḳāf	pronounced by the tongue and the velum palati.
k	ك	Kāf	
l	ل	Lām	
m	م	Mīm	
n	ن	Nūn	
h	ه	Hā	
w	و	Wāw	
y	ى	Yā	

At the beginning of words and syllables the *Alif* (') is not represented. The termination of feminine nouns (*at*) is represented by *ah*, except where a genitive follows. The case terminations (*nom. u; gen. i; acc. a*) and their nasalized forms (*un; in, an*) are not represented. The article is invariably transcribed *al*; no account is taken of the assimilation of the *l* to a following consonant. The vowels are used in their so-called Continental pronunciation.

Argo point.   One of the three commonly detectable points along the vertical circle through the sun at which the degree of polarization of diffuse sky radiation goes to zero.

Arcturus.   See Alpha Bootis.

areal velocity.   In celestial mechanics, the area swept out by the radius vector (a line from the central body or focus to the orbiting body) per unit time. The areal velocity is constant for a central force.

areography.   The physical geography of Mars. The study of the surface features of Mars. The tradition of different terms for different planets is dying out and "geography" is acceptable.

argon.   A chemically inert gaseous element.

argument.   In astronomy, an angle or arc, as in argument of perigee.

argument of latitude.   In celestial mechanics, the angular distance measured in the orbit plane from the ascending node to the orbiting object; the sum of the argument of perigee and the true anomaly.

argument of perigee.   In celestial mechanics, the angle or arc as seen from a focus of an elliptical orbit from the ascending node to the closest approach of the orbiting body to the focus. The angle is measured in the orbital plane in the direction of motion of the orbiting body.

Ariel.   Second satellite of Uranus orbiting at a mean distance of 192,000 km.

Aries (The Ram).   A zodiacal constellation widely known in ancient times. Its importance lies in the fact that in ancient times the sign of Aries marked the vernal equinox and the beginning of spring. Today precession has moved the vernal equinox to Pisces.

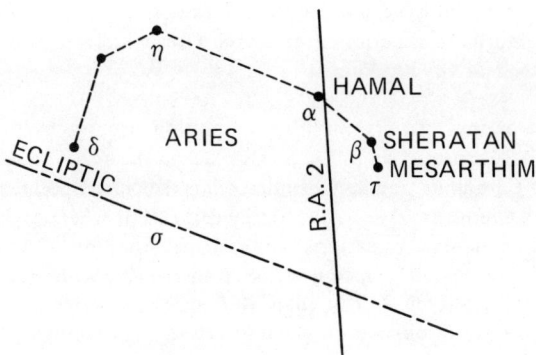

Aries (Astrology).   Aries is the first sign of the Zodiac, because in ancient times when astrology originated the vernal equinox was in Aries. It has the sign of the Ram and runs from March 21 to April 20 each year. It was thought to govern mental activity.

Aries, First Point of.   The vernal equinox. The point at which the sun crosses the celestial equator from south to north. Right ascension and celestial longitude are both measured from this point. The point has moved gradually away to Pisces from the constellation Aries (where it was when early astronomers determined its position) owing to the Precession of the Equinoxes.

Aristarchus.   A Greek astronomer of the first half of the third century B.C. He made estimates of the distances of the Sun and Moon and held that the earth not only rotated on its axis, but moved around the Sun.

artificial satellite.   Any artificial device orbiting around a planet or satellite.

ascending node.   That point at which a planet, planetoid, or comet crosses to the north side of the ecliptic plane; that point at which a satellite crosses to the north side of the equatorial plane of its primary.

ashen light.   A faint glow on the unilluminated side of Venus named from a fancied resemblance to faint glowing coals.

aspects.   The apparent position of celestial bodies relative to one another; particularly, the apparent positions of the Moon or a planet relative to the Sun. See also conjunction, elongation.

aspects of the inner planets.   Because the inferior planets, Mercury and Venus, revolve faster than the Earth, they gain on the Earth and therefore appear to oscillate to the east and west with respect to the Sun's place in the sky. Their aspects are unlike those of the Moon, which has all values of elongation up to 180°. After passing superior conjunction beyond the Sun, the interior planet emerges to the east of the Sun as an evening star and slowly moves out to greatest eastern elongation. Here it turns west and, apparently moving more rapidly, passes between us and the Sun and inferior conjunction into the morning sky. Turning east again at greatest western elongation, it returns to superior conjunction. Greatest elongation does not exceed 28° for Mercury and 48° for Venus.

aspects of the moon.   In its monthly revolution around us the Moon moves continuously eastward relative to the Sun's place in the sky. The moon's elongation at a particular time is its angular distance from the Sun. Special positions receive distinctive names and are known as the aspects of the Moon. When the Moon overtakes the Sun, generally passing it a little to the north or south, the elongation is not far from 0°. The moon is in conjunction with the Sun when the two bodies have the same celestial longitude. It is in quadrature when its elongation is 90° either east or west. The moon is in opposition when its celestial longitude differs by 180° from that of the Sun so that its elongation is not far from 180°.

aspects of the planets.   Aspects of the planets relative to the Sun are similarly reckoned to the aspects of the Moon. For the conjunctions of the planets with the Moon and with one another, however, the predictions in the Diary of the American Ephemeris and Nautical Almanac are the times to the nearest hour when the two bodies have the same right ascension.

aspects of the superior planets.   Because the superior planets revolve more slowly than the Earth does, they move eastward in the sky more slowly than the Sun appears to do, so that at intervals they are overtaken and passed by the Sun. With respect to the Sun's position they seem to move westward and to attain all values of elongation 0° to 180°. The aspects of the superior planets are the same as those of the Moon.

association.   A group of stars fairly close together in space, moving together and probably formed together. Generally, an association is less tightly bound than an open cluster. It has long been known that O and early B stars of the Milky Way and also T Tauri-type variables tend to occur in groups less compact than open clusters. Soviet astronomer V. A. Ambartsumiam called attention to the temporary existence of such associations. Too feebly bound by gravitation to hold together very long, their stars are evidently so youthful that they have not had time to disperse. They are sometimes classed as OB associations and T associations, depending on which type of stars are dominant.

A stars.   Stars of spectral class A defined by spectral properties indicating effective surface temperatures of about 7400 to 9900° K.

astatic.   Without orientation or directional characteristics.

asterism.   A group of stars which may or may not form a constellation. The stars of the Big Dipper, for example, form an asterism in the constellation of the Great Bear.

asteroids (the minor planets).   Observationally, any small interplanetary body not identified as a planet. Asteroids are a few kilometers in diameter or less. The largest is about 1000 km across. Thousands of asteroids orbit the Sun between the orbits of Mars and Jupiter, and others are known elsewhere in the solar system. Most are fragments of bodies ranging up to at least 500 km in diameter, and are sources of many meteorites. They range in composition from carbon-rich rocky bodies to metal-rich rocky bodies, according to spectroscopic evidence. Asteroids are numbered in order according to determinations of their orbits, and named according to the whims of their discoverers.

asteroids, Amor.   A group of asteroids with orbits that cross Mars' orbit (as projected on the ecliptic plane) but do not cross Earth's orbit. Typical Amor orbits reach from the asteroid belt to a point between Earth and Mars. The group is named after the prototype asteroid, Amor.

asteroids, Apollo.   A group of asteroids with orbits that cross Earth's orbit (as projected on the ecliptic plane). Collisions with Earth could occur but since the orbits generally cross above or below the ecliptic plane collisions with Earth are unlikely in most cases for hundreds of millions of years. About 20 have been observed carefully and have known orbits, but others have been discovered and subsequently lost. Spectral evidence indicates they are rocky bodies a few kilometers across. Most meteorites may be fragments broken off from them.

asteroids, Trojan.   Two groups of asteroids in Jupiter's orbit about 60° ahead of and behind Jupiter. Existence of such bodies was predicted in effect by the mathematician J. L. Lagrange when he found that small bodies could occupy these positions stably in any system of two massive bodies in circular orbit around each other. Lagrange pointed out that Jupiter and the Sun form such a system. The first Trojan discovered was Achille, found in 1906. The largest is Hektor, an elongated body roughly 300 km long. By tradition they are named after Homeric heroes of the Trojan wars.

astigmatism (telescope).   A distortion of an optical resulting from the failure of rays from different parts of the same concentric zone of the objective to focus in the same place. Generally causes stellar images to be elongated, and circular objects to be imaged as ellipses.

Astraea.   An asteroid with a diameter of about 60 miles.

astro.   A prefix meaning star or stars and, by extension, sometimes used as the equivalent of celestial, as in astronautics.

astroballistics.   See astrodynamics.

astrobiology.   The study of extraterrestrial life or protobiological extraterrestrial chemistry.

astrochemistry.   The study of the chemical nature of celestial bodies.

astro compass.   An instrument used to determine direction by sighting heavenly bodies of known position.

astrodynamics.   A branch of dynamics that treats of the motions of celestial bodies and of space vehicles, and of the forces acting on them in motion.

astrolabe.   An instrument for the measurement of the altitude of a celestial object.

astrology.   The assumption, without any physical basis, that the positions of planets and other bodies as seen from the earth affect the lives of different people in different ways. Astrological practices grew in popularity around 1000 B.C. and probably existed earlier. Ancient records and modern records in primitive societies show that astronomical events and sociological events were often tabulated together (i.e., the sighting of a comet and the death of a king three years later). Astrology apparently evolved from the belief that the first event caused the second. Elaborate and often inconsistent systems have been devised to predict human events or tendencies based on planetary, lunar, and solar positions. A record showing the positions of these bodies in the sky at any given place and moment is called a horoscope, and is said to be the most important of these devices. The horoscope corresponding to a person's birth is said to affect his life. Many studies have listed numerous examples in which astrological predictions were incorrect, but the superficial resemblence of complexity in interlocking astronomical cycles and the ups and downs of human life apparently encourage continued belief in astrology among many people.

astrometry.   The study of positions, distances, and motions of stars (distinguished from studies of stellar physics, chemistry, and evolution).

astronaut.   A human traveling in space.

astronautics.   The science and technology of travel in space by either manned or unmanned vehicles. Broadly, the term includes rocket technology, orbital dynamics, life support technology, and related fields.

astronomical calender.   See ephemeris.

astronomical ephemeris.   See ephemeris.

astronomical symbols.   Symbols denoting astronomical phenomena, in many cases derived from ancient astrological practice.

☿	Mercury		
♀	Venus		
⊕	Earth		
♂	Mars	☌	Conjunction. Where two bodies have the same right ascension but not necessarily the same declination.
♃	Jupiter		
♄	Saturn		
♅	Uranus	☍	Opposition. Where two bodies differ 180° in right ascension.
♆	Neptune		
♇	Pluto	▢	Quadrature. Where two bodies differ 90° in right ascension.
☉	Sun		
●	New Moon		
☽	First Quarter		
○	Full Moon		
☾	Last Quarter		

astronomical twilight.   Period during which the Sun is between 0° and 18° below horizon.

astronomical unit.   A unit of distance equal to that of the geometrical mean distance of the Earth from the Sun which is 1.496×10^{13} cm, 92.9×10^6 miles or 8.3 minutes.

astronomy.   The study of physical matter and energy in the universe, including its evolution and origin.

astronomy, radio.   The study of long-wave celestial sources. Radio waves from space were found in 1931. These waves are part of the electromagnetic spectrum but are not detectable with ordinary telescopes. They are received by antennas that are similar to the antennas of radio and television sets. Radio sources include planets, stars, nebulae, and diffuse interstellar gas.

astrophotography.   The photographic imaging of celestial objects and the technology and development of photosensitive materials designed to meet certain problems of detection such as imaging at certain wavelengths.

astrophysics.   A branch of astronomy that treats of the physical properties of celestial bodies such as luminosity, size, mass, density, temperature, and chemical composition.

atmosphere.   The outer, gaseous layers of a star (or planet).

atmosphere (Earth).   The Earth's atmosphere is a mixture of gases surrounding the earth's surface to a height of several hundred kilometers. From its average pressure of 1.013×10^6 dynes/cm^2 at sea level, the mass of the entire atmosphere is calculated to be 5.2×10^{21} grams, or somewhat less than one millionth the mass of the Earth itself. The lower atmosphere is divided into two layers; the troposphere and the stratosphere.

COMPOSITION OF EARTH'S ATMOSPHERE

Gas		Fraction of Dry Air	
Name	Symbol	by volume (%)	by weight (%)
Nitrogen	N_2	78.0	75.5
Oxygen	O_2	20.9	23.1
Water	H_2O	0.1–28	0.06–1.7
Argon	A	0.93	1.3
Carbon dioxide	CO_2	0.033	0.50
Neon	Ne	0.0018	0.0013
Helium	He	0.00052	0.000072
Methane	CH_4	0.00015	0.00008
Krypton	Kr	0.00011	0.0003
Carbon monoxide	CO	0.000006–0.0001	0.000006–0.0001
Sulfur dioxide	SO_2	0.0001	0.0002

Note: Some constituents vary with height or geographic location such as CO_2, CO, SO_2. Others are greater in industrial areas.

atmosphere (escape).   The ability of a celestial body to retain an atmosphere around it depends on the velocity of escape from the body at the top of the atmosphere, and on the atmospheric temperature. A molecule moving upward at the top of the atmosphere faster than escape velocity will overcome the pull of gravity and escape into space. The higher the temperature, the faster molecules will move. Thus atmospheres tend to escape from low-gravity, low-mass planets, especially if the upper atmosphere is heated. Calculations show that a celestial body will lose half its atmosphere in only a few weeks if the typical molecules move $1/3$ as fast as the escape velocity in that atmosphere. The required time is increased to a few thousand years if the factor is $1/4$ and to a hundred million years if the factor is $1/5$.

atmosphere (ionosphere).   The ionosphere is the region of any planet's atmosphere that is most affected by the impact of high-frequency radiations and high-speed particles from outside, ionizing atoms and molecules. On earth, the ionosphere is often regarded as the layer from altitude 70 to 300 km, where the ionized gases are more abundant, but the designation may be extended to include the entire upper atmosphere.

atmosphere (stratosphere).   A middle portion of Earth's atmosphere from 12 to 50 km above the surface. More generally, the lowest portion of any planet's atmosphere in which temperature increases or stays nearly constant with height. In Earth's stratosphere, currents are horizontal and only a little water vapor is present.

atmosphere (troposphere).   The lowest turbulent part of the Earth's atmosphere, about 0 to 12 km above the surface. More generally, the lowest part of any planet's atmosphere in which the temperature decreases in height. Its chief constituents are nitrogen and oxygen (in the proportions of 4 parts to 1 by volume), carbon dioxide, and water vapor; there are other gases in relatively small amounts and dust in variable quantities. By their strong absorption of infrared radiations, water vapor carbon dioxide absorb infrared heat radiated from the surface, creating a temperature gradient and preventing rapid escape of heat from the ground.

atmospheric extinction.   Reduction in intensity of light passing through the atmosphere by absorption and scattering as photons interact with gas and dust.

atmospheric pressure.   The pressure at any point in an atmosphere due to the weight of the atmospheric gases above the point concerned.

atmospheric radiation.   Radiation emitted within a planet's atmosphere, usually of infrared or radio wavelength.

atmospheric refraction.   Refraction resulting when a ray of radiant energy passes obliquely through an atmosphere. It may be called astronomical refraction if the ray enters the atmosphere from outer space. It is most pronounced near the horizon, where rays have the longest atmospheric path; it distorts the appearance of the rising and setting sun and moon.

atmospheric shell.   Any one of a number of strata or layer of a planet's or star's atmosphere.

atmospheric stability.   See atmosphere (escape).

atmospheric windows.   Portions of the spectrum (wavelength intervals) in which the atmosphere is relatively transparent to radiation. At other wavelengths, light may be blocked by absorption due to certain molecules such as H_2O.

atoms.   Atoms are the building blocks of all materials. They consist of a nucleus composed essentially of protons and neutrons and a surrounding group of orbiting electrons. The electron is the lightest of these major constituents; its mass is 9.1055×10^{-28} grams and it carries unit negative charge of electricity. The proton is 1836.57 times as massive as the electron and carries unit positive charge. The neutron has about the same mass as the proton and is electrically neutral. The nucleus of the atom ranges progressively from the single proton of the ordinary hydrogen atom to compact groups of protons and neutrons in the heavier atoms. Each added proton contributes one unit to the positive charge in the nucleus. In the normal atom the nucleus is surrounded by negatively charged electrons equal in number to the protons, so that the atom as a whole is electrically neutral. However, one or more electrons can be temporarily gained or lost, in which case there is a net charge and the atom is called an ion.

atomic clock.   A time-keeping device based on a constant frequency of radiation of a chosen atomic species; atomic clocks now provide the most accurate known time standards.

atomic mass.   The mass of a neutral atom or atomic particle, usually expressed in atomic mass units. In this system the mass of a proton is 1.007 amu. Atomic mass is sometimes called atomic weight. 1 amu $= 1.6605 \times 10^{-24}$ grams.

atomic number.   The atomic number of an element is the number of protons in the nucleus and also the number of electrons around the nucleus of the neutral atom. All atoms having the same atomic number are defined as being the same chemical elements.

atomic particle.   One of the particles of which an atom is constituted, as an electron, neutron, or proton. Smaller particles exist and are sometimes called subatomic particles.

atomic structure.   The arrangement of electrons in an atom.

atomic weight.   Atomic mass.

A.U.   Abbreviation for astronomical unit.

A.U.R.A.   Association of Universities for Research in Astronomy, an organization which jointly administers Kitt Peak National Observatory.

Auriga (The Charioteer). A large constellation of ancient origin to which Ptolemy assigned fourteen stars. Its area is 657 sq. deg. The principal star is the brilliant Capella with solar-type spectrum. Auriga contains 9 open clusters of which the 3 brightest are fine objects in the Messier list. It is pictured as a benevolent celestial giant, seated on the Milky Way, a region glowing with the light of myriads of far-off telescopic stars.

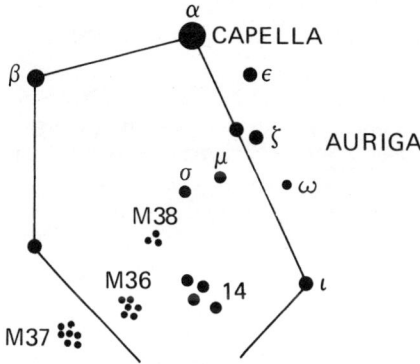

aurora. Glowing arcs or bands seen intermittently at high latitudes in the night sky. Colors include white, yellow, red, and green. Often consisting of ghostly, swirling curtains moving slowly across the whole sky, it originates at a height of about 90 km (a sharp lower limit) to 300 km, and sometimes reaches 700 km. It is caused by atomic particles that collide with the upper atmosphere and excite atoms which emit light as they return to their normal, or ground states. The particles originate on the Sun and travel to Earth in about one day. On reaching the earth they may be trapped in the Earth's magnetic field in so-called Van Allen belts, before being "dumped" into the atmosphere near the north and south magnetic poles. The resulting glows are called Aurora Borealis or Northern Lights, and Aurora Australis or Southern Lights, respectively.

australite. Type of tektite found in Australia.

autumn. Period from autumnal equinox to winter solstice.

autumnal equinox. The time (about September 23) when the overhead sun crosses the equator on its apparent migration from north to south. On this date and the date of spring equinox the length of day and night is approximately the same in all latitudes.

averted vision. Vision not with the central retina but with the more light-sensitive outer part of the retina. Averted vision is achieved by looking a few degrees to one side of the object being contemplated. This makes fainter objects such as stars more visible than they would be by direct vision.

axis. A straight line about which a body rotates.

azimuth.   Angular distance around the horizon measured from the north point toward the east. Due east, for example, has an azimuth of 90°. In astronomy, the azimuth of a celestial object is given by the azimuth of a horizon point directly below the object (i.e., azimuth of the intersection of the horizon and the great circle containing the zenith and the object). For example, an azimuth of an object 45° up on the southern meridian would be 180°.

azimuthal mounting.   A mounting allowing an instrument to swing in the horizontal, or azimuthal plane.

azimuthal projection.   Map projection made so that distances of lines radiating from the central point are of same proportional lengths as corresponding lines on the sphere.

B

Babinet point. One of the three commonly detectable points of zero polarization of diffuse sky radiation—neutral points—lying along the vertical circle through the Sun. The other two are the Arago point, and Brewster point.

back focal length (telescope). The distance from the back surface of a lens to the focal point.

backward scatter. The scattering of radiant energy into the hemisphere of space bounded by a plane normal to the direction of the incident radiation and lying on the same side as the incident ray; the opposite of formal scatter.

Bailly. A large crater (about 290 km) on the visible side of the Moon. Although it was once listed as the largest crater on the Moon, larger craters or basins have been recognized in the 1960's.

Baily's beads. The string of light patches seen when the Moon eclipses the Sun, just before and just after the moment of totality and caused by the penetration of the last rays of sunlight between the irregularities of the Moon's surface. So-called after Francis Baily, an English astronomer who described them in 1836.

Balmer discontinuity. The apparent abrupt change in spectral brightness at the blueward end of the Balmer series at 3646 Å.

Balmer lines. A series of lines in the spectrum of atomic hydrogen, starting with H-alpha at wavelength 6563 Å, H-beta at 4861 Å, H-gamma at 4341 Å, H-delta at 4102 Å. The lines crowd together and end abruptly at the Balmer discontinuity at 3646 Å. J. J. Balmer, a Swiss physicist, derived in 1885 an empirical formula by which the wavelength of any line in the series can be calculated.

band spectrum. An emission or absorption spectrum consisting of a number of bands, each band being a very large number of closely spaced lines which overlap on low-dispersion spectograms.

bands. Spectral absorption or emission feature given by two or more atoms combined to form a molecule. Band spectra usually appear as a long series of closely packed lines extending from the "head" toward the red or violet. Cooler stars display broad bands as well as lines in their spectra. In such stars the collisions of atoms and molecules are not excessively violent and a few kinds of two-atom molecules are bound together strongly enough so that they can survive the buffeting without being dissociated into individual atoms. Just as atoms can emit or absorb light by changing energy states, so can molecules. But here the number of permitted energy changes is enormous, for not only may the electrons in the individual atoms make transitions, but the two atoms of a molecule may vibrate, for instance, toward and away from each other, and rotate around each other with various energies. These energy changes in a molecule can create a system of very closely spaced lines which appear as bands. The bands of molecules such as CN, CO, C_2, and TiO are prominent in the spectra of some of the cooler stars.

bar.   A unit of pressure equal to 16^6 dyne per square centimeter (10^6 barye), 1000 millibars, or 29.53 inches of mercury. One bar is the approximate sea level atmosphere pressure on earth.

Barlow lens.   A small lens with a concave surface and a negative focal length which is placed between the objective and the eyepiece. The result is an effective increase in the focal length of the objective, increasing the magnification. Such a lens is most useful for visual observation because it increases the range of magnifications attainable with a given set of eyepieces.

Barnard's star.   The largest known proper motion is that of a telescopic star called "Barnard's star" after the name of the astronomer who first noticed its swift apparent motion. This 10th magnitude star in Ophiuchus is moving with respect to the background stars at the rate of 10″.3 a year, so that it advances in 175 years through an angle equal to the Moon's apparent diameter. If all the stars were moving as fast as this and at random, the forms of the constellation would be altered appreciably in the course of a lifetime. Barnard's star has also received attention as a result of several astrometric analysis indications that it is attended by one or more companions of about Jupiter's mass. These have been claimed as indication of another planetary system; other analysts have questioned the observations.

barometric wave.   Any wave in the atmospheric pressure field. The term is usually reserved for short-period variations not associated with cyclonic-scale motions or atmospheric tides.

barosphere.   The atmosphere below the critical level of escape.

Barr scale (telescope).   A scale used to measure the longitudinal motion of the knife edge in a Foucault testing device.

barred spirals.   Galaxies are divided into three groups, elliptical, spiral and irregular. There are two main types of spiral galaxies, normal and barred. A minority of spirals have a bright bar that slices across the nucleus. The arms begin at the two ends of the bar and wind outward. The normal spiral galaxies consist of a central region (nucleus) to which a number of spiral arms appear to be attached. The barred spirals differ in the respect that the arms seem to start from a luminous bar that crosses the nucleus. Below are types of galaxies proposed by Edwin P. Hubble which have been modified in minor respects but the general classification is substantially unchanged. (See Fig., p. 29.)

barycenter.   The center of gravity of any two-body system such as the earth and moon, earth and sun, or a double star.

barycentric oscillation.   A small, oscillatory motion of the Earth around its orbit about the Sun owing to the fact that the center of mass of the Earth–Moon system does not coincide with the center of the Earth.

SPIRALS

ELLIPTICAL

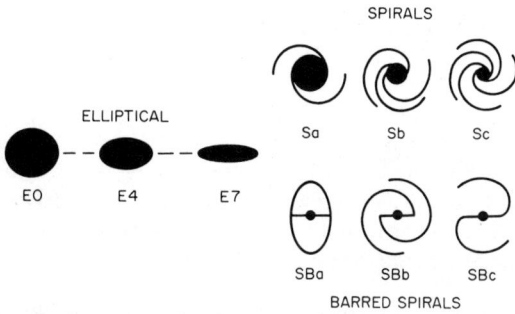

EO   E4   E7   Sa   Sb   Sc

SBa   SBb   SBc

BARRED SPIRALS

basalt. A broad class of erupted volcanic lava rock composed mostly of feldspar minerals and having little quartz. It is usually dark grey or grey-brown in color. Basalts are an important planetary rock type, forming lunar maria (q.v.), much of Earth's crust, and probably much of the volcanic flows of Mars.

basic. A rock type poorer in silica and richer in iron than other types.

Bayer's Uranometria. The star maps of Bayer's Uranometria (1603) introduced the present plan of designating the bright stars of each constellation by small letters of the Greek alphabet. In a general way, the stars are lettered in order of brightness and the Roman alphabet is drawn upon for further letters. The full name of a star in the Bayer system is the letter followed by the possessive of the Latin name for the constellation. When several stars in the constellation have nearly the same brightness, they are lettered in order of their positions in the mythological figure, beginning at the head. Thus the seven stars of the Great Dipper, not much different in brightness, are lettered in order of position.

BD (followed by a number). The Bonner Durchmusterung (catalogue) identification number of a star.

Bellatrix. *See* Gamma Orionis.

B emission stars. They are luminous and blue and appear to be surrounded by non-expanding gaseous rings or shells. In these stars we do not see emission lines with the violet absorption edges that characterize a swelling nebula. Instead, the typical Balmer line for B emission stars is a very wide absorption line, sitting atop of which is a narrower emission line containing a central narrow absorption line.

Benetnasch. The star at the end of the handle of the Big Dipper. Also called Alcaid. Eta Ursa Majoris.

beral coating. An alloy of beryllium and aluminum used to provide a reflective coating on a mirror.

Bessel, F. W. German astronomer who first discovered stellar parallax in 1838.

Besselian star numbers. Constants used in the reduction of a mean position of a star to an apparent position.

Beta Andromedae.   Known as Mirach, an Arabian name meaning "girdle." Its visual magnitude is 2.0. It is estimated at 78 light-years distant and slowly moving farther away. It is a red giant.

Beta Centauri.   A prominent southern star of visual magnitude 0.6. It is a hot, bluish star of spectral type B at an estimated distance of 420 light-years. Known as Hadar to the Arabs, it makes a prominent pair with the nearby star Alpha Centauri. Alpha and Beta are sometimes called "the southern pointers" because a line projected through them points to the Southern Cross.

Beta Cygni.   The foot of the Northern Cross, which is also the "Beak of the Swan," is marked by Beta. It was called Albireo by the Arabs, meaning "the Hen's Beak." One of the most beautiful doubles in the sky, it is especially interesting because the striking contrast in the color of the components (one is blue, the other, reddish) is visible in a small telescope of a few centimeters aperture.

beta decay.   Radioactive transformation of an atom in which the atomic number changes by one with the emission of an electron, and the mass number remains unchanged.

Beta Geminorum.   Pollux, a K giant of visual magnitude 1.15. Estimated distance is about 36 light-years, making it the nearest giant star.

Beta Leonis.   Beta Leonis is called Denebola, the name meaning "Lion's Tail." It is a main sequence A star of visual magnitude about 2.14. Its estimated distance is about 43 light-years. In astrological superstition, it was considered unlucky.

Beta Lyrae.   A remarkable, much-studied eclipsing binary with two unequal minima in each full period. The range is 3.38 to 4.36 and the full period is 12.91 days. One member of the pair may be enveloped in a large gas cloud.

Beta Orionis.   Better known as Rigel, a brilliant star of magnitude 0.1, sparkling like a bluish white diamond in the upraised foot of Orion. Rigel has the distinction of being one of the most luminous stars in all the heavens, a hot supergiant of spectral type B8. Its diameter is about 35 times that of the Sun. Its light requires 460 light-years to make the journey.

Beta Tauri.   A blue giant star of magnitude 1.6 and distance 180 light-years.

Betelgeuse.   See Alpha Orionis.

bias.   A source of error in observations. It may be due to an unconscious expectation on the part of the observer or a design problem in the instruments.

Bielid meteors.   A November meteor shower closely identified with Biela's comet. In years past there have been wonderful displays of these meteors, most notably in 1872 and 1885.

big bang.   A hypothetical event approximately 16 ± 4 billion years ago, in which all the matter in the present universe began expanding from an extremely compressed, high-density state with virtually infinite initial temperature. The theory of the big bang is supported by observations such as the mutual recession of the galaxies, and especially the weak 3K black-body radiation permeating the universe, discovered after its prediction by the big-bang theory.

binary.   A system of two co-orbiting stars.

binaries, astrometric.   Binary systems detected by orbital motions around one another.

binaries, eclipsing.   Binary systems detected by eclipses when one component passes in front of the other as seen from the Earth. This means that the total light received from the system is reduced, and the star seems to give a long, slow "wink." The most famous is Algol in the northern constellation of Perseus.

binaries, spectroscopic.   Binaries detected by the Doppler shifts of their spectral lines.

binaries, visual.   Binary stars that can be separated (resolved) with the telescope.

binoculars.   A pair of telescopes optically equivalent and arranged parallel for use with both eyes.

biosphere.   The near-surface zone between the Earth's interior and atmosphere within which forms of terrestrial life are commonly found; the outer portion of the geosphere and inner and lower portion of the atmosphere.

birefringent filter.   A form of monochromator, consisting of a series of alternating layers of plate cut in a special manner from a birefringent (doubly-refracting) crystal (i.e., quartz or calcite), and of films of a polarizing material, such as Poloroid.

bistatic radar.   In the bistatic radar method, the transmitted signal is sent out from a station on Earth, but the echo is received by simple, lightweight equipment carried on a vehicle close to the reflecting surface (i.e., Venus). The echo signal is then amplified and transmitted to Earth. The signal received on the ground is thus much stronger than an echo obtained directly from the target.

black body.   An ideal emitter which radiates energy at the maximum possible rate per unit area at each wavelength for any given temperature. A black body also absorbs all the radiant energy in the near visible spectrum incident upon it.

black dwarfs.   Objects that do not have sufficient mass to start a self-sustaining conversion of hydrogen to helium and so fail to become true stars. Thus, any object of less than about 0.085 solar mass. In some usages, planets are considered black dwarfs. Other usage considers black dwarfs as more massive than Jupiter (i.e., about 0.001 to 0.085 solar mass). The binary L726-8 may contain an example.

black hole. Classically, an object of such extremely high density and high surface gravity that no radiation can escape. Theoretically, black holes range from stellar masses of kilometric scale to subatomic particles. Recent use includes atomic-scale examples in which limited radiation, such as gamma rays, may escape. Black holes may form following the explosive mass-loss and collapse of evolved super-massive stars.

blink comparator. An instrument which enables two photographic plates of the same region taken by the same telescope but at different times to be examined alternately in rapid succession. The plates are placed side by side on a stage and illuminated alternately several times a minute. The observer sees coincident images of first one plate, then the other. This rapidly changing presentation of the two plates to the eye facilitates the discovery of moving objects, variable stars and novae which attract the observer's attention.

blinking. Examination of images in a blink comparator.

Bode's law. An empirical law giving the approximate distances of the planets from the Sun. The law is: To each of the series of numbers 0, 3, 6, 12, 24, 48, 96, 192, 384, (doubled each step after the first), add 4. We then obtain the series 4, 7, 10, 16, 28, 52, 100, 196 and 388 which when divided by 10 represents approximately the distances of the planets from the Sun in astronomical units.

Bode's law and the mean distances of the planets.

Planet	Bode's Law	Actual Distance
Mercury	4 + 0 = 0.4	0.39
Venus	4 + 3 = 0.7	0.72
Earth	4 + 6 = 1.0	1.00
Mars	4 + 12 = 1.6	1.52
Asteroids	4 + 24 = 2.8	2.8
Jupiter	4 + 48 = 5.2	5.20
Saturn	4 + 96 = 10.0	9.54
Uranus	4 + 192 = 19.6	19.19
Neptune	—	30.06
Pluto	4 + 384 = 38.8	39.44

Bohr's atomic model. In 1913 the Danish physicist Niels Bohr proposed a model for hydrogen, the simplest of all atoms, which consisted of a single electron revolving around a positive proton, like a planet moving around the Sun. The model brilliantly accounted for many features of hydrogen's spectrum. Bohr followed with models like miniature solar systems for more complex atoms, in which negative electrons revolve around the positive nucleus, but these models failed significantly for complex atoms.

bolide. A large meteor, generally one that explodes with a noise; a fireball. (Greek: *bolis*, "missile").

bolometric luminosity. The total energy, including energy at all wavelengths in the electromagnetic spectrum, radiated per second by a star.

bolometric magnitude. The magnitude of a star based on its entire energy output at all wavelengths in the electromagnetic spectrum. Some energy receivers, such as thermocouple, bolometer, and radiometer, respond to light of any color. They allow an estimate of bolometric magnitude after correction for energy absorbed in the Earth's atmosphere.

Boltzmann's constant. Constant of proportionality relating an atom or molecule's mean kinetic energy to the gas temperature: $K = 1.38 \times 10^{-16}$ *ergs/Kelvin*.

Boltzmann equation. An equation giving the distribution of particles among energy states for a group of the atoms, ions, and electrons in a gas of known temperature.

bond albedo. The fraction of incident light reflected in all directions from a sphere exposed to parallel light.

Bootes (the Herdsmen). An easily recognizable constellation dominated by the yellow-orange 1st magnitude star Arcturus forming a distinctive equilateral triangle with Spica and Denebola.

Bose-Einstein statistics. Theoretical statistics of properties of subatomic particles based on certain assumptions about their quantum–mechanical properties.

boundary value problem. A physical problem specified by rates of change of variables (a differential equation) and certain "boundary conditions" such as the initial conditions in a system of particles or the conditions after a certain elapsed time.

Boyle's law. The finding that when a gas is compressed and kept at a constant temperature, its pressure times its volume remains constant ($PV = C$).

Brackett series. Hydrogen spectral lines associated with hydrogen's 4th energy level.

bremsstrahlung. Electromagnetic radiation produced by the rapid change in the velocity of an electron or another fast, charged particle as it approaches an atomic nucleus and is deflected by it. (German, meaning "braking radiation"). Sometimes called free–free radiations since the electron remains unbound to any atom before and after the radiation process.

Brewster point. One of the three commonly detectable points of zero polarization of diffuse sky radiation along the vertical circle through the Sun; the other two are the Argo point and the Babinet point.

brightness temperature. The temperature of a black body radiating the same amount of energy per unit area as the observed body at the specified wavelength.

Busch lemniscate. The locus in the sky of all points at which the plane of polarization of diffuse sky radiation is inclined 45° to the vertical.

C

C star. See carbon star.

Caelum (The Sculptor's Tool). A dim, dull constellation adjoining Lepus, Eridanus, and Columba. There is no star above magnitude 4.

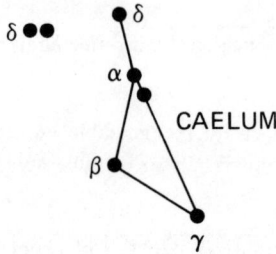

calendar, Gregorian. The modern western calendar. It was proposed by astronomers under Pope Gregory to correct the errors that had accumulated in the Julian calendar. Toward the end of the 16th century the vernal equinox had retreated from March 21 to March 11. To restore the equinox to Mar. 21, ten calendar days were suppressed from the year 1582. In that year, the day following October 4 became October 15 for those adopting the Gregorian system. So that the calendar would remain in step with the seasons, the average length of the calendar year was adjusted to more closely approximate the tropical year by making common years of those century years having numbers not evenly divisible by 400.

calendar, Julian. The calendar prepared under Julius Caesar. Its chief feature was the adoption of 365 1/4 days as the average length of the calendar day, accomplished conveniently by the plan of leap years. Three common years of 365 days were followed by a fourth year containing 366 days (thus the leap year in our era has a number evenly divisible by four). Since the calendar began with Spring equinox in March, the leap year was added at the end of the year—at the end of February. In lengthening the calendar from the 355 days of the old lunisolar plan to the common year of 365 days, Caesar distributed the additional 10 days among the months. With further changes made in the reign of Augustus, the months assumed their present lengths. After Caesar's death in 44 B.C. the 5th month, Quintilis, was renamed July in honor of the founder of the new calendar. The 6th month, Sextilis, was later renamed August in honor of Augustus.

calendar, lunar. The simplest and earliest calendar used by many societies. Each month began originally with the "new moon," the first monthly appearance of the crescent moon after sunset. Long controlled only by observation of the crescent, this calendar eventually operated by fixed rules. In the fixed lunar calendar the 12 months of the lunar year are alternately 30 and 29 days long, making 354 days in all and thus having no fixed relation to the year of the seasons. The Mohammedan calendar is a survivor of this type.

calendar, lunisolar. The lunisolar calendar tries to keep in step with both the moon's phases and the seasons. It began by occasionally adding a 13th month to the short lunar calendar year to round out the year of the seasons. The extra month was later inserted by fixed rules. The Jewish calendar is the principal survivor of this type.

calendar, Roman. The early Roman calendar dates formally from the traditional founding of Rome in 753 B.C. It was originally a lunar calendar of a sort, beginning in the spring and having 10 months: March, April, May, June, Quintiles, Sextilis, September, October, November and December. For many centuries thereafter the years were counted from 753 B.C. and were designated A.U.C. Two months, January and February, were added later and were eventually placed at the beginning so that the numbered months have since then appeared in the calendar out of their proper order. In the 12-month form the Roman calendar was of the lunisolar type.

calendar, solar. The solar calendar makes the calendar year conform as nearly as possible to the year of the seasons and neglects the moon's phases; its 12 months are generally longer than the lunar months. Only a few early nations, notably the Egyptians and eventually the Romans, adopted this type of calendar, now in wide use.

callipic cycle. Four metonic cycles, or 76 years.

Callisto. The second largest satellite of Jupiter orbiting at a mean distance of 1,883,000 km. Its diameter is about 4900 km. Also called Jupiter IV.

Camelopardalis (the Giraffe). A large, extremely faint constellation containing no stars above the 4th magnitude.

canali. The Italian name for the line-like markings reported on Mars in 1877 by G. V. Schiaparelli. In Italian "canali" signifies natural waterways or "channels."

canals. Dark, controversial, line- or streak-like features on Mars. Though streaky features were sketched by earlier observers, true "canals" were first reported in 1877 by G. V. Schiaparelli and described in great detail in following decades by Percival Lowell. Spacecraft observations indicate the reported linear markings were essentially illusory, though patchy alignments and streaks due to dust deposits do exist.

Cancer (the Crab). The 4th constellation of the ancient Zodiac, to which Ptolemy assigned 13 stars. Although this group contains no star brighter than 4th magnitude, it is easily found by its close proximity to Leo. Cancer is noted for its two bright clusters, M.44 and M.67. M.44 is known as Praesepe. (See Fig. p. 36.)

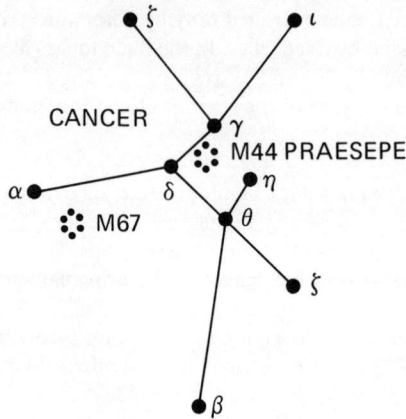

candela. The unit of luminous intensity in the International System of Units, defined in 1967 as equal to the luminous intensity of $1/600,000$ m^2 of a black body at the temperature of freezing platinum and a pressure of $101,325$ newtons/m^2.

Canes Venatici (the Hunting Dogs). Most of the stars of Canes Venatici were once included in Ursa Major. Canes Venatici lies between Ursa Major and Bootes, the area of the modern constellation being approximately 465 sq deg. It contains a region rich in distant galaxies and the beautiful globular cluster M.3.

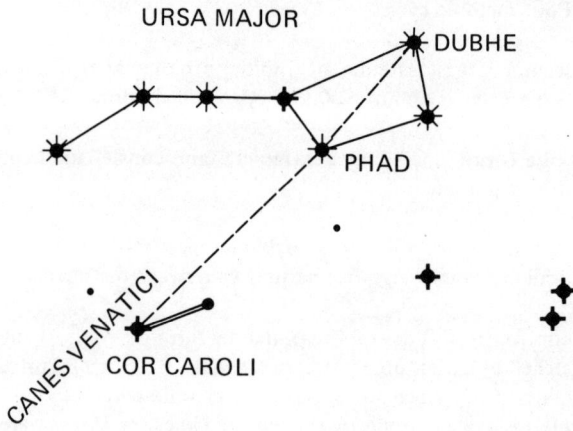

Canis Major (The Greater Dog). A conspicuous constellation lying mainly just south of the Milky Way between Orion and the long train of bright stars which formerly composed Argo Navis. It is one of the Ptolemaic groups to which were assigned 18 stars. Dominated by the brilliant white Sirius, the modern area is about 380 sq deg. This constellation is rich in fine objects; several double stars are particularly beautiful. (See Fig. p. 37.)

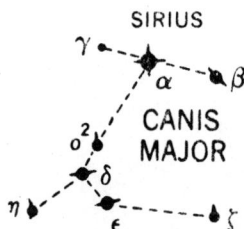

Canis Minor (the Little Dog). The constellation of Canis Minor, the small dog, contains only 183 sq deg. Known to ancients as the companion of the larger animal Canis Major, it is easily recognized by the two bright stars assigned to it. The brighter of these is Procyon, a brilliant yellow star.

Canopus. A brilliant southern star in the constellation Argo. It is the second brightest star in apparent magnitude (second only to Sirius), and one of the most luminous supergiants. Its apparent magnitude is $-.86$. Canopus is accompanied by a much smaller star of magnitude 3.5, and the two revolve about a common center of mass in 6 3/4 days. The pair is about 180 light-years away.

Canopus sensor. A device on many space vehicles designed to locate and track the isolated bright star Canopus in order to maintain the vehicle's orientation.

Capella. A star of zero magnitude in the constellation Gemini. It is one of the half dozen brightest stars in the sky. It is the leader of Auriga, the celestial charioteer. It has a yellowish color and a face temperature about the same as that of the Sun, but whereas the Sun is a main-sequence star, Capella is a triple system with two yellowish giants and a small red dwarf. It lies at a distance of 42 light-years.

Capricornus (The Sea Goat). The goat is the 10th constellation of the Zodiac. Ptolemy assigned 28 stars to this group which is not conspicuous but easily recognized by the two stars χ and β following Sagittarius. The modern area is 414 sq deg. There are some interesting double stars in Capricornus and the globular clusters M.30 and M.72. (See Fig., p. 38.)

carbon cycle. A six-stage series of thermonuclear reactions beginning with the capture of a hydrogen nucleus or proton by the nucleus of a carbon atom. The complete process converts hydrogen to helium and releases energy. The carbon cycle was suggested as a stellar energy source by H. Bethe and C. F. von Weizsacker. In its successive reactions the carbon nucleus $_6C^{12}$ captures protons to form increasingly more complex

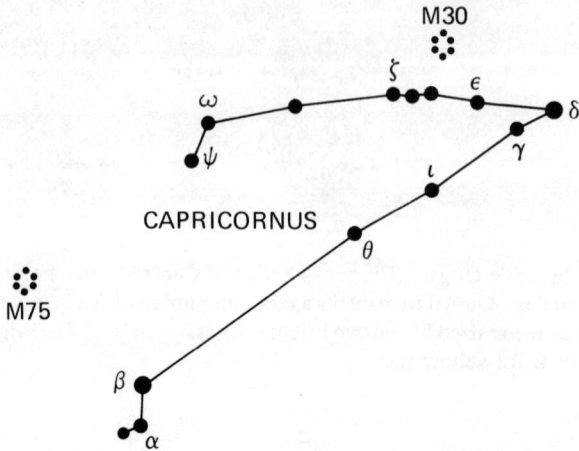

nuclei until finally, after 4 protons in turn have been captured, the product splits into a helium nucleus and a carbon nucleus like the one that started the chain. This carbon atom can serve to start the process over again. The reaction is therefore called a cycle, in which carbon plays the role like that of a chemical catalyst. Sometimes called the CN cycle. The cycle can be written:

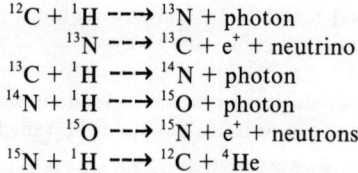

$$^{12}\text{C} + {}^1\text{H} \dashrightarrow {}^{13}\text{N} + \text{photon}$$
$$^{13}\text{N} \dashrightarrow {}^{13}\text{C} + e^+ + \text{neutrino}$$
$$^{13}\text{C} + {}^1\text{H} \dashrightarrow {}^{14}\text{N} + \text{photon}$$
$$^{14}\text{N} + {}^1\text{H} \dashrightarrow {}^{15}\text{O} + \text{photon}$$
$$^{15}\text{O} \dashrightarrow {}^{15}\text{N} + e^+ + \text{neutrons}$$
$$^{15}\text{N} + {}^1\text{H} \dashrightarrow {}^{12}\text{C} + {}^4\text{He}$$

carbon stars.   A rare class of spectral types (q.v.) R, N and S, unusually rich in carbon or carbon compounds such as C_2 and CN, which exist in their atmospheres.

carbonaceous chondrites.   Black, weak chondrites containing roughly 2 to 4 percent of carbon, partly in the form of the element, but mainly as complex organic compounds. Some are weak enough to be crumbled by hand. They comprise an estimated 3–5% of all meteorites seen to fall. They are important for being the least-altered material known to date from the solar system's origin.

Carina (The Keel).   An extended constellation of area 494 sq deg formed by Gould in 1879 from the southern portion of the very large ancient group of Argo Navis, to which Ptolemy had assigned 45 stars in all. The Milky Way passes through the region; it contains many bright stars including the second brightest, Canopus. Carina is very rich in objects for small telescopes. There is a fine series of double stars and open clusters. (See Fig. p. 39.)

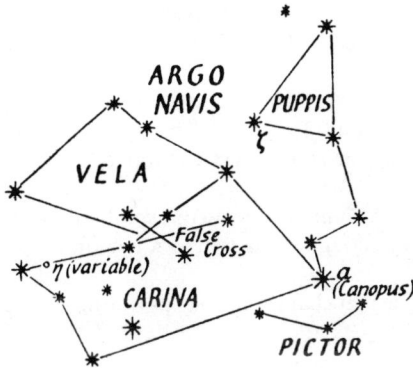

Cartesian coordinates. A coordination system in which the location of points in space are expressed by reference to three perpendicular planes called coordinate planes.

Cassegrain telescope. A reflecting telescope in which the light is reflected from the (parabolic) main mirror onto a convex (hyperbolic) secondary mirror and back through a hole in the center of the main mirror, thence into the eyepiece.

Cassegrainian focus. The focal point in a Cassegrain telescope.

Cassini's division. The dark space or gap between the outer ring of Saturn and the bright inner ring. First seen by G. D. Cassini in 1675, it is probably caused by perturbation of ring particles by Saturn's satellites.

Cassiopeia. This fine constellation with its 6 principal stars in the form of a flat distorted M or W lies in and north of a rich part of the Milky Way. It is one of the oldest and best known of the star groups, to which Ptolemy assigned 13 stars. The modern area is 598 sq deg. There are some fine double stars in Cassiopeia and roughly 50 open clusters. No globular clusters have been found and all the known planetary nebulae are very faint.

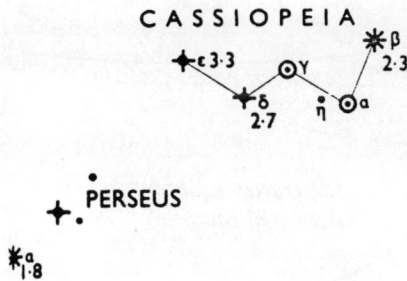

Cassiopeia A (3C 461).   The strongest extrasolar radio source, believed to be the remnant of a supernova observed in 1667 and lying roughly 3 kpc away.

cataclysmic variable stars.   Variables with irregular light variations including novae, recurrent novae, and flare stars. They may all be close binarys in which eruptive activity occurs when gas from one star is shed and dumped on the other.

catadioptric telescope.   A compound telescope combining optics of refractor and reflector. It employs a correcting lens as well as a mirror.

catalyst.   Any substance that changes the rate of a chemical reaction without itself being changed.

CD (followed by a number).   The Cordaba Catalogue number, similar to the BD number but referring to stars in the southern sky.

celestial body.   Any object observed in the sky.

celestial equator.   The great circle of the celestial sphere located halfway between the north and south celestial poles. It is the plane of the Earth's equator projected onto the sky.

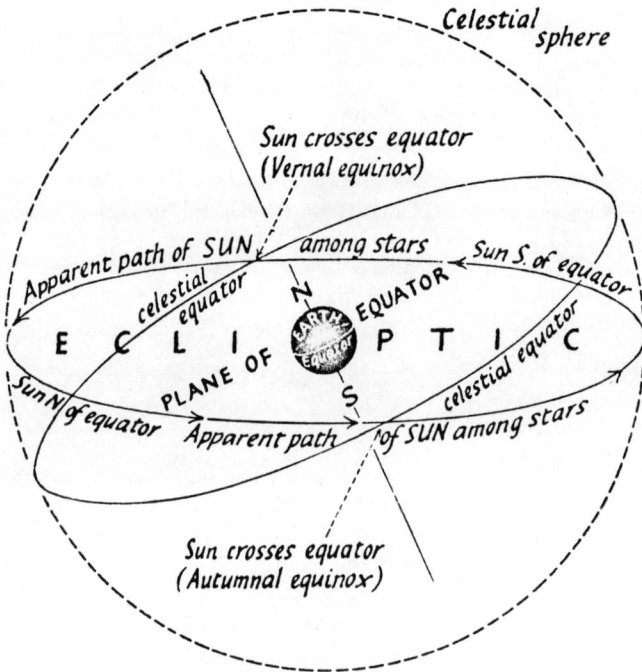

celestial horizon.   The great circle on the celestial sphere located halfway between the zenith and nadir and 90° from each. This is the horizon of astronomy as distinguished from the actual visible horizon, which is the frequently irregular line where the Earth and the sky seem to meet.

celestial latitude.   Angle north or south of the ecliptic to the object.

celestial longitude.   Angle measured eastward along ecliptic from the vernal equinox to foot of the celestial latitude circle passing through the object.

celestial meridian.   The great circle that passes through the zenith, the nadir, and the north and south points of the horizon. All circles that pass through the zenith and nadir are called vertical circles.

celestial sphere.   The "sphere of the sky" with moon, stars, and planets "attached." Half of it appears over us on a clear dark night, and maps of the sky are maps of this sphere. The position of a star on the celestial sphere indicates a direction in space. Its center is generally taken as the observer's position. The apparent westward movement of the Sun and other bodies across the sky, as though the celestial sphere were rotating daily around the Earth from east to west, is called diurnal motion of the heavens and is an effect of the Earth's rotation on its axis from west to east.

The celestial sphere.

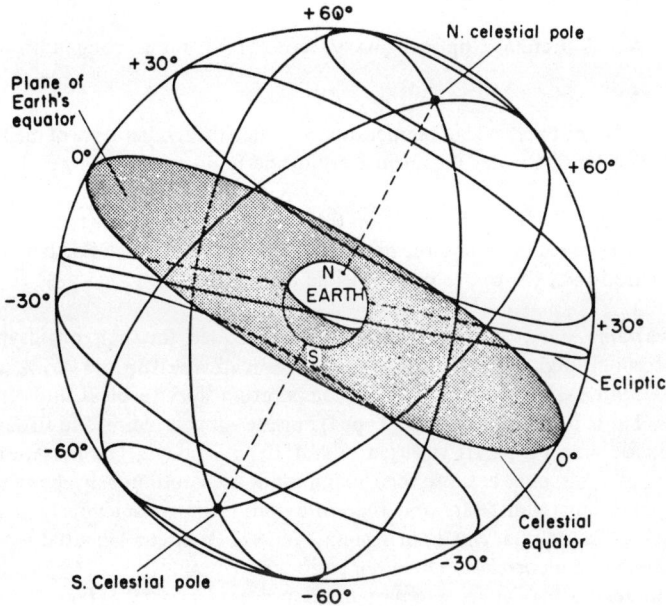

Centaurus (The Centaur). A very large southern constellation included in the Almagest of Ptolemy in A.D. 150 in which he assigned 37 stars. At present the two brightest of these can hardly be seen above the horizon at latitudes 30°N (2,000 years ago they were about 10° farther from the south pole). The brightest member, Alpha Centauri, is the dominant star of the closest known star system (in this case involving three co-orbiting members). The closest known star, Proxima Centauri, is a faint member of this system (magnitude 11). Centaurus is a conspicuous constellation with an unusually large number of bright stars. It contains a wide range of interesting double stars and some attractive open clusters.

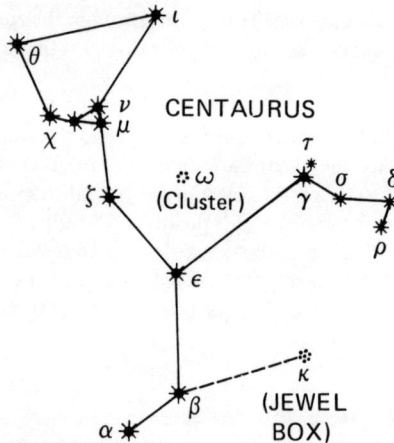

Centaurus A. A peculiar elliptical galaxy (NGC 5128); it is a strong radio and x-ray source.

centigrade. A uniform scale of temperature in which the temperature of melting ice is 0° and that of boiling water (at normal atmospheric pressure) is 100°.

central meridian. An imaginary line on the Sun, Moon, or a planet, drawn through the body's poles of rotation, and the point in the planet from which Earth would be directly overhead (i.e., the center of the visible disk).

cepheid variable stars. Cepheid variable stars, named for their prototype Delta Cephei, are single high luminosity stars that periodically swell up and shrink with consequent periodic variations in light, color, temperature, spectral class, and other characteristics. Those in the galactic disk (Type I) pulsate with periods of 5 to 10 days; those in the galactic halo (Type II) with periods of 10 to 30 days. The pulsations make cepheids easy to discover because their magnitudes are continuously changing. They are giant and supergiant stars and therefore stand out conspicuously among the brighter stars of a globular cluster or a galaxy. In 1912 Henrietta Leavitt discovered an extremely important correlation between periods and luminosities of Cepheids, allowing astronomers to use Cepheids to measure distances of galaxies as far as 3 mpc away.

Cepheus. An inconspicuous constellation of area 588 sq deg representing the kingly father of Andromeda; it is an ancient star to which Ptolemy assigned 11 stars, the brightest of which is only magnitude 2.6. About half the constellation is immersed in the Milky Way.

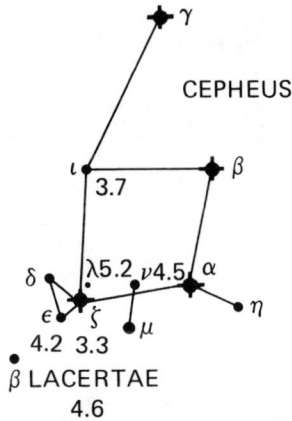

Cerenkov radiation. The radiation from a charged particle, whose velocity is greater than the velocity of light in the medium where the particle is traveling. The particle will continue to lose energy by Cerenkov radiation until its velocity is less than the medium's speed of light.

Ceres. The largest of the asteroids, with an estimated diameter of 1000 km, a sidereal period of 4.6 years, an orbital semimajor axis of 2.8 A.U., and a magnitude at opposition of 7.4.

cerium oxide (telescope). A polishing agent made from the rare earth element cerium.

Cetus (The Whale). Fourth in order of size among the constellations with an area of 1,231 sq deg, it is one of the ancient groups to which Ptolemy assigned 22 stars. It lies mainly just south of the equator immmediately following Aquarius in a rather barren part of the sky and contains many double stars including the variable Mira with its fine spectrum, but no open or globular clusters. (See Fig., p. 44.)

Chamaeleon (The Chamaeleon). This small constellation occupying 132 sq deg between Carina and the south polar Octans, was proposed by Bayer in 1604. It consists of a few scattered stars, some of about magnitude 4 and some fainter. There is little of telescopic interest in this constellation.

Chandrasekhar limit. A theoretical maximum possible mass for a white dwarf star, equal to 1.44 solar masses under certain assumptions; derived by the Indian–American astrophysicist S. Chandrasekhar.

μ

ζ

α
2·5

γ
3·6   δ 4·0

• α Piscium
3·9

• MIRA

θ 3·8

C E T U S

ζ
3·9   X

η 3·5

τ

υ

β

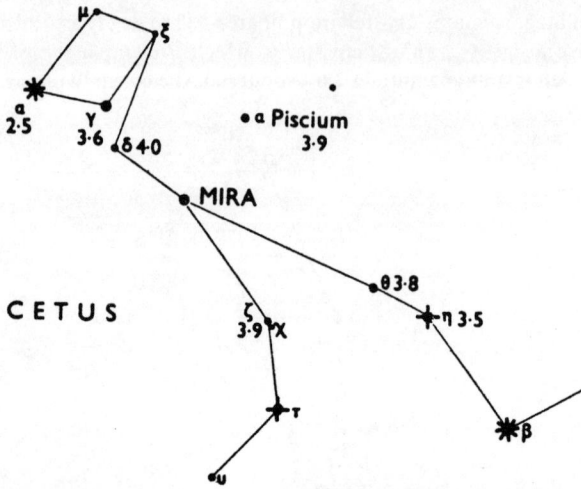

charge.   An electrical property of matter, either positive or negative, in units of the charge on proton (+) or electron (−).

Chiron.   An object with semimajor axis 13.7 A.U., inclination 6. °9, eccentricity 0.38, and period 50.7 years, discovered in 1977 by Charles Kowal at Hale Observatory. It is believed to be about 100 to 700 km in diameter. Its orbit overlaps Saturn's and reaches nearly to Uranus. Its nature is something of a puzzle, since it is much larger than known comets but has an orbit different from any known asteroids. It may be an example of a new category of objects populating the outer solar system. It may be more clearly observed when it reaches perihelion in 1996.

chondrites.   Over 90 percent of the stony meteorites have small spheroidal inclusions, a millimeter or two in diameter called chondrules (Greek: *chondros*, "grain of wheat"). These meteorites are referred to as chondrites; those that do not include chondrules are known as achondrites. Chondrules are composed of magnesium-iron silicates. In chondrite meteorites, chondrules are imbedded in a matrix consisting of irregular masses, mainly of magnesium-iron silicates—olivine and pyroxene in particular—together with pieces of broken chondrules.

chondrules.   See chondrites.

chromatic aberration.   The defect of an uncorrected lens or system of lenses in failing to bring light of all colors to one and the same focus. Blue light is focused closer to the objective lens than red light. It occurs only in the refracting (lens) type of telescope and can be mostly eliminated by making the lens of two different kinds of glass.

chromosphere.   A layer of the Sun's atmosphere that lies just above the photosphere, or visible surface, and below the corona. The Sun's chromosphere is not a uniform atmospheric layer. It consists of many fine, luminous columns of gas, including spicules, which appear side by side like blades of grass, rising and falling and rapidly changing in intensity. The lower chromosphere is mostly neutral hydrogen gas at about 7500° K; the upper chromosphere contains ionized hydrogen at temperatures as high as one million degrees.

Circinus (The Compasses).   The pair of compasses is a small southern constellation of area 93 sq deg. Constituted in 1752 by Lacaille, it lies in the Milky Way just south of and following the pointers to the Southern cross, forming a right angle with them. There are 3 open clusters.

cirrocumulus (clouds).   A type of cloud often composed of small, white, flaky masses or very small, shadeless, globular masses arranged in groups or lines, or (more often) in ripples resembling those of the sand in the sea shore. This form results from an unstable atmospheric condition surrounding a sheet of cirrostratus or cirrus.

cirrostratus (clouds).   According to the International classification, "a thin, whitish veil, which does not blur the outlines of the Sun or Moon, but gives rise to halos." These halos are due either to the refraction of light through ice crystals or to the mere reflection of light by the faces of the crystals. These halos are positive proof that cirrostratus clouds, like the cirrus from which they frequently develop, often consist of crystals. Like the cirrus, cirrostratus is formed by air forced to the highest levels in areas surrounding low barometric pressures.

cirrus (clouds).   The Latin meaning of "curl" or "wrap" is descriptive of these icy and fibrous cloud forms. "Detached" clouds of delicate, fibrous, often silky appearance, without shading and generally white in color, cirrus clouds also appear in long parallel bands or in banks which cross the sky like meridian lines and appear to converge to a point in the horizon.

cislunar.   Relating to the space between the Earth and the Moon, or between the Earth and the Moon's orbit.

cis-planetary.   Relating to the space between the Earth's orbit and the orbit of a given planet.

civil time.   Identical with local mean solar time, with zero defined as midnight and 12^h as noon. Hence the hour angle of the mean sun plus twelve hours defines civil time.

clepsydra.   An early device for showing time based on the uniform rate of flow of water, oil, or sand through a small hole or pipe.

closed universe.   (1) A hypothetical model of the universe in which the galaxies are not expanding at escape velocity; the density of material in the universe creates enough gravity to cause the galaxies to slow down and eventually fall back together. (2) A hypothetical model of the universe with curved space and finite volume.

closest approach.   The event that occurs when two planets or other celestial bodies are nearest to each other as they orbit about the Sun or other primary.

cloud attenuation.   Usually, the reduction in intensity of microwave radiation by clouds in the Earth's atmosphere. At centimeter wavelength, clouds produce Rayleigh scattering. The attenuation for both ice and water clouds is due largely to scattering rather than to absorption.

cloud types.   Four major families are defined, three according to their levels, and a fourth penetrating all cloud altitudes:
  Family A—High clouds, stratified
  Family B—Medium clouds, stratified
  Family C—Low clouds, stratified
  Family D—Vertically developed clouds
Within these families, there is a total of ten primary types recognizable by appearance. See cloud types such as cirrus, cumulus, etc.

clusters.   Clusters are small groups of stars more or less close together. Clusters are of two major types: globular and open. Most clusters (and other objects such as nebulae and galaxies) are designated by NGC (New General Catalogue) numbers; if the object is included in Messier's catalogue, M numbers may be added. Examples of prominent clusters include:

Cluster Name	Constellation	Distance	Diameter	Type
Pleiades (M45)	Taurus	127 psc	4 psc	Open
Hyades	Taurus	42	5	Open
NGC 1039 (M34)	Perseus	440	4	Open
NGC 5139 (ω Cen)	Centauri	5000	20	Globular
NGC 5272 (M3)	Canes Venatici	13000	13	Globular
NGC 6205 (M13)	Hercules	7700	11	Globular

clusters, galactic.   Old term for open star clusters, derived from the clusters' typical location near the galactic plane.

clusters, globular.   A massive, nearly spheroidal group of stars, usually ranging from 10,000 to one million solar masses in total mass. Globular clusters are believed to have formed about 12 to 13 billion years ago. Most are located above and below the galactic plane and form a halo around the galactic disk. Their stars are nearly pure hydrogen and helium and contain fewer heavy elements than the younger stars (such as the Sun) in the galactic disk.

clusters, moving.   The few clusters that are located close enough to us so that their proper motions can be detected. Measurements of their motions can be converted into measures of their distances to provide a check on other methods of distance measurement.

cluster of galaxies. A group of associated galaxies, usually within 10 to 100 galaxy diameters of each other.

cluster, open. A loose grouping of young stars (ages typically one million to a few hundred million years) usually located in the galactic spiral arms. Masses are, typically, a few hundred solar masses.

CN cycle. See carbon cycle.

Coal Sack. A dark nebula in the Southern Milky Way where stars are greatly obscured by a cloud of dark material. It is about 550 light-years away.

codeclination. Ninety degrees minus the declination.

coherent radiation. Radiation in which the waves are in phase with each other. In normal starlight, in which the various waves originate independently in different ions, atoms and molecules, the phases are randomly distributed and the radiation is incoherent.

colatitude. Ninety degrees minus the latitude.

collimate (telescope). To render rays of light parallel and aligned with the telescope axis.

collimating lens. A lens which converts rays of light from a point source into parallel rays.

collimation error. The angular difference in direction between two nominally parallel lines of sight; specifically the angle by which the line of sight of an optical or radio instrument differs from what it should be.

collision. An encounter between two particles that changes their existing momentum and energy conditions.

collision broadening. In spectroscopy, the broadening (in wavelength) of an emission line due to the interruption of the radiating process by a collision of the radiator with another particle. Lines originating in stars' atmospheres are often broadened.

color excess. The amount of reddening of distant stars that is produced by the scattering of their light by interstellar material.

color index of a star. (1) The photographic magnitude minus the visual magnitude. Because the magnitude scale is such that smaller numbers denote brighter stars, color index is positive for all stars redder than Sirius and negative for the bluer stars. A star's color index depends on the temperature of its surface. The index provides a useful measure of temperature as long as passage through interstellar dust has not reddened the starlight. It is a simple tool for statistical studies of stars and of the structure of the Milky Way. (2) Since detectors sensitive to many different wavelengths have come into use in recent years, color index has been broadened to mean any difference in magnitude measured at two wavelengths—defined as shorter-wave magnitude minus longer-wave magnitude.

color magnitude diagram. A plot of apparent brightness or luminosity of stars (in magnitudes) against their colors. It is another version of the H–R diagram.

color temperature. An estimate of the temperature of an incandescent body, determined by observing the wavelength at which it is emitting with peak intensity (its color), and using that wavelength in Wien's Law. The temperature to which a blackbody radiator must be raised in order that the light it emits may match a given light source in color.

Columba (The Dove). First adopted as a constellation in Bayer's atlas of 1603. A small group of area 270 sq deg lying immediately south of Lepus and recognized by 2 stars of magnitude 3 and several less bright ones. Its telescopic objects include a number of double stars and fine bright clusters.

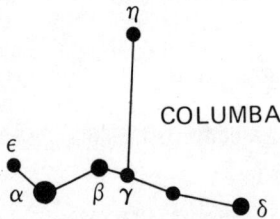

coma. (1) The defect of a lens or lens system in which star images appear triangular or comet-shaped. (2) The hazy, bright region round the nucleus of a comet.

Coma Berenices (Berenices Hair). The star group known as Coma Berenices (the hair of Berenices) was known to the ancients but included either with Leo or with Virgo to the south. Tycho Brahe in 1602 catalogued it separately as a constellation. The modern area is 386 sq deg; there are no conspicuous stars. Coma is rich in telescopic objects including many distant galaxies.

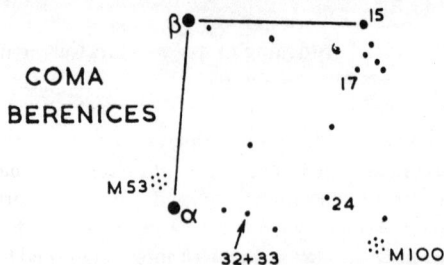

combination coefficient. A measure of the specific rate of reaction of small ions due to either (a) union with neutral particles to form new large ions or (b) union with large ions of opposite sign to form neutral particles.

comet. An object consisting of a bright stellar-looking nucleus surrounded by an extensive halo and tail of dust particles and gases (such as H_2, CN, C_2, OH) traveling on an elliptical orbit around the Sun. Many have orbits extending from the region of the terrestrial planets far beyond Pluto. Nuclei of comets are believed to be "dirty icebergs"—icy bodies consisting of frozen H_2O, CH_4, NH_3, and related substances with dust particles distributed through the icy matrix. They are believed to have formed in the earliest days of the solar system and been preserved in a swarm beyond Pluto (the Oort cloud) where they are too far from the Sun's heat to be destroyed. Upon passage through the inner solar system, they are heated by the Sun, causing release of gases and dust and sometimes fragmentation of the comet nucleus into two or more bodies.

comet Arend-Roland. First spotted in November of 1956 by two Belgian astronomers, Comet Arend-Roland went past the Sun during April of the following year. After that, the comet assumed a strange appearance with two tails pointing in opposite directions, one pointing away from the Sun and the other toward the Sun. Careful studies of photographs later showed that the sunward pointing spike was part of a fan-shaped tail stretching beyond the comet in the plane of its orbit and seen only in projection as if on the sunward side. A few such phenomena have been seen before, with the sunward-pointing spikes occurring after perihelion and lasting only a few days.

comet Biela. This was a small, periodic comet and a member of Jupiter's family. It had a short period of 6.6 years. It was seen to return several times before its famous passage of 1846. In November of that year, it appeared as usual. A month later it became pear-shaped and then divided into two comets, each with a short tail. In the predicted return years of 1859 and 1866, they were no longer visible. In 1872, which would have been the next apparition year, a dazzling meteor shower was seen. This cometary debris continued to circulate in the same orbit for a time, and then disappeared.

comet, Brooks (Comet 1911V). One of the many comets discovered by Brooks, this comet is known principally for its great beauty. It also underwent a number of major tail developments. In July, it had no tail at all. Toward the end of September it had "grown" a slender tail; during October the tail formed sheetlike streams, and then great numbers of delicate streamers enhanced this already stunning object in the night sky.

comet, De Cheseaux's (The Great Comet of 1744). Appearing in the heavens in the spring of 1744, De Cheseaux's comet was unique for its fanlike six tails. All six were reported visible to the naked eye.

comet designation. A comet is designated provisionally by the name of its discoverer and the years of its discovery, followed by a letter indicating the order in which the discovery is announced; (i.e., Comet 1956h). The later permanent designation is the year of perihelion passage (not always the year of discovery) followed by a Roman numeral in order of perihelion passage during that year (i.e., Comet 1957 II). Many comets, especially the more remarkable ones, remain permanently known by the name of the discoverer or discoverers.

comet, Donati's (Comet 1858VI).   Donati's comet was one of the most striking comets seen during the nineteenth century. Discovered in June 2, 1858, it was visible to the naked eye for 112 days in that year; it was watched with telescopes for more than 9 months. Its long, elliptical orbit reached about 10 times the distance of Neptune. Its period is about 2,000 years. It is possible that the bright comet mentioned by the Roman writer Seneca in 146 B.C. was Donati's comet. In the telescope, this was one of the most beautiful of all comets. Its tail was some 50 million miles long, curved wide at the end, and split at times into 2 or 3 streamers. The head passed in front of the star Arcturus, which was easily seen through the comet. Donati's comet was of particular interest because of the series of gaseous envelopes expelled from its nucleus. Eventually this famous comet threw off 7 envelopes in just a few days.

comet Encke.   This comet had the shortest orbiting period, 3.3 years. With eccentricity 0.85, inclination 12°, and perihelion distance approximately that of Mercury, this object has been observed on dozens of returns to perihelion.

comet, Great Comet of 1811.   This immense comet attracted so much attention and caused so much excitement that a vintage of wine was named after it. Its great head was considerably larger than the Sun. A long period comet, the Great Comet of 1811 may take some 3,000 years to return. It was discovered in France in March of 1811, and was observed for as long as 17 months—a record for those prephotographic days.

comet, Great Daylight Comet of 1910.   Not to be confused with the returning Halley's comet of that year, this comet's claim to fame is that it was fully visible in broad daylight very near the Sun, and for a few days was much brighter than Venus. It was discovered by three railroad workers in Africa who spotted it with their naked eyes. It passed perihelion on January 17 and later developed a forked tail. For a day or two, this was probably the brightest comet seen since the beginning of the twentieth century.

comet, Halley's.   This famous comet, the first known periodic comet, is named in honor of Edmund Halley who predicted its return. Halley calculated the parabola orbit of the bright comet of 1682 and noted its close resemblance to the orbits that he had similarly calculated for earlier comets of 1531 and 1607 from records of their places in the sky. Concluding that they were appearances of the same comet, which must therefore be moving in an ellipse, Halley predicted that it would return again in 1758. The comet was sighted that year according to prediction; it returned again in 1835 and 1910. Halley's comet is the only conspicuous comet having a period less than 100 years. The revolution is retrograde. (See Fig., p. 51.)

comet head.   The nucleus and the coma together are usually referred to as the head of the comet. While the heads of comets average an immense diameter of 800,000 miles, the actual nuclei are very small by comparison.

Halley's comet

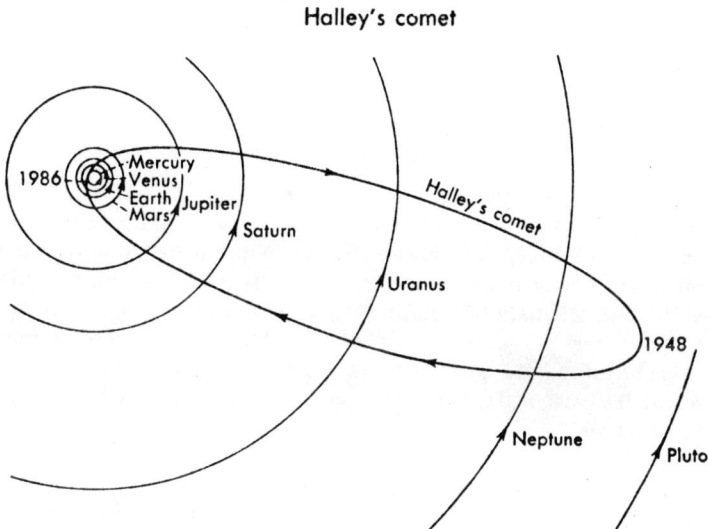

comet, Jupiter's Family. Several dozen comets have orbits closely related to the orbit of Jupiter. In each case the aphelion or one of the nodes is near the orbit of Jupiter so that the comets often pass close to the planet itself. These are members of Jupiter's family of comets. Their close approaches to the great planet result in such great orbital perturbations that the configuration of the family is not stable. The periods of revolution of these comets around the Sun are mostly between 5 and 9 years, averaging a little more than half of Jupiter's period. The orbits are generally not much inclined to the ecliptic and the revolutions are all direct. These comets are believed to have been temporarily captured into their present orbits by close approaches to Jupiter.

comet, Kohoutek. A widely publicized comet observed in 1973. After its discovery many months before maximum brightness, some comet astronomers predicted it would be extremely bright, perhaps visible in daylight around Christmas, a prediction sensationalized in the press. (Other astronomers, reviewing behavior of similar types of comets, predicted that it would not be very prominent, but these predictions were less publicized.) The comet was a great disappointment to the public when it failed to brighten to the degree expected, but it was widely observed scientifically and led to new data on cometary physics.

comet, long-period. Comets can be roughly divided into two groups. By far the most numerous comets belong to the group with long periods of revolution, often up to thousands of years. Called long-period comets, they have orbits reaching far beyond the solar system. Sometimes they have aphelions thousands of astronomical units from the Sun. The period of such a comet is so long in terms of human existence that only one appearance of its passage near Earth has ever been recorded.

comet, P/Encke.   The 2nd periodic comet to be "discovered" by analysis, by John Encke, later director of the Berlin Observatory. Encke compared the comets observed in 1786, 1795 and 1818, and realized they were the same body. The period of Encke's comet is much shorter than that of most, and since its predicted appearance in 1822 it has been observed at every return.

comet, Schwassmann-Wachmann (Comet 1925 II).   The short period comet is one of the most remarkable known. It has an unusually circular orbit for a comet, traveling entirely between the orbits of Jupiter and Saturn. With a period of 16 years it was the first comet that could be observed in every part of its orbit, from aphelion to perihelion. This strange celestial body is also subject to astonishing changes in brightness, perhaps due to irregular escape of gases. At times it flares up and becomes more than 500 times as bright as it was just a few days previously. Unfortunately, it is always so far from both the Earth and the Sun that even at its brightest a good telescope is necessary to see it at all.

comet, short-period.   Comets which have revolution periods of not more than a few hundred years, and orbits generally not extending beyond the solar system. There are approximately 100 of these, 85 possess long elliptical orbits that lie entirely within the orbit of the planet Neptune. Six of these do not revolve in the regular direction—west to east as seen against the stars—but have retrograde motion, orbiting the Sun from east to west. About 60 of these short-period comets have periods under 12 years, and almost all of the 60 have aphelions near the orbit of the giant planet Jupiter.

comet, spectrum.   The spectrum of a comet is characterized by bright bands produced by gases set glowing by the Sun's radiation. Prominent gases are carbon (C_2), methyne (CH), hydroxyl (OH), ammonia radicals (NH_2 and NH), and cyanogen (CN). These are rather unstable and are soon transformed into more durable molecules, such as carbon monoxide, carbon dioxide, and nitrogen, as they are driven from the coma into the tail. Hydrogen has more recently been discovered among comet gases. The unstable constituents of the coma are formed by action of sunlight on parent molecules of methane, ammonia, and water in the nucleus. These materials remain frozen when the comet is far from the Sun, and begin to evaporate as it approaches the Sun.

comet, sun-grazers.   Comets that pass unusually close to the Sun, often causing disruption. When Comet 1947n passed within about 9 million miles of the Sun, its nucleus was split in two parts. During its famous perihelion passage of 1896, Biela's comet became pear-shaped and then divided into two comets, each with a short tail. Comet Brooks was followed by four fragments, after its close encounter with Jupiter in 1886. All of these are examples of comet groups, or sets of comets with nearly identical orbits near the Sun.

comet's tail.   The tail of a comet is composed of gas and fine dust. It develops as the comet approaches the Sun and is likely to become conspicuous if the perihelion is close to the Sun. Tail generally points directly away from the Sun, repelled by a force exceeding that of the Sun's attraction. The repulsive force is usually ascribed to the pressure of the Sun's radiation, perhaps increased irregularly by collision with streams of high-speed particles emerging from the Sun. There are two major classes of comet tails: Type I streams out almost directly opposite to the Sun, and Type II curves around backward in the orbit plane. The Type I tails away from the nucleus showing emission bands spectra of ionized molecules and are called ion tails. The Type II tails show sunlight scattered from dust particles and are called dust tails. Diagram shows orientation of tail as comet moves around sun.

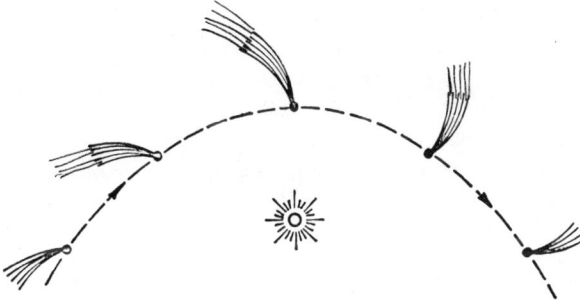

comet, Tycho Brahe's (The Comet of 1577). During 1577, the great Danish astronomer Tycho Brahe observed this comet with great care and made astronomical history. Previously it had been thought that comets were simply vaporous phenomena in the Earth's atmosphere, possibly a part of the remote heavens which were then regarded as fixed, immovable, possibly subject to shattering by a body traveling beyond the atmosphere. Yet even without a telescope, Brahe was able to place limits on the comet's parallax, proving that it was not an atmospheric phenomenon at all and that it was at least three times farther away from the Earth than the Earth from the Moon. Thus, Tycho Brahe received the credit for proving once and for all that the ancient doctrine of the immutability of the heavens was false.

companion.   The smaller body in a co-orbiting star or planet binary.

comparison star.   A star of known constant brightness against which a variable object is compared.

complement.   An angle equal to 90° minus a given angle.

condensation.   The physical process by which a vapor becomes a liquid or solid.

condensation nucleus.   A particle, either liquid or solid, upon which condensation of vapor begins.

condensation sequence.   The sequence in which various mineral grains appear by condensing as a nebular gas (often around a new star) cools from about 2500 K to 100 K. First appear refractory minerals such as aluminum oxide, followed by metallic grains, silicates (rocky particles), and lastly ices.

conformal projection.   A map projection on which all small or elementary figures upon the surface of the Earth (or other body) retain true shape, the meridians and parallels being at right angles to one another.

conics.   The conics or conic sections are the ellipse, parabola, and hyperbola. They are sections cut from a circular cone, as shown. These same curves are the paths followed by one body in the gravitational field of another, depending on the relative velocity.

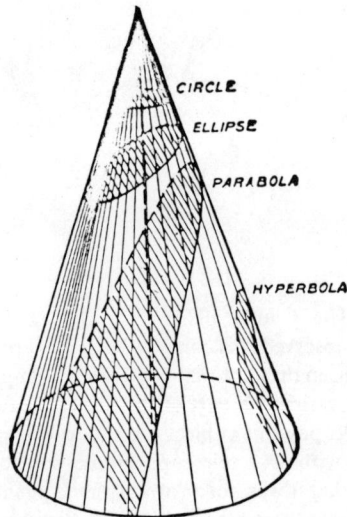

CIRCLE
ELLIPSE
PARABOLA
HYPERBOLA

The Conics.

conic projection.   A map projection which can be imagined as drawn on the surface of a cone. The meridians appear as straight lines along which the parallels as concentric circles may be spaced in such a way as to give some desired qualities such as conformality or equal area.

conjunction.   (1) The moment when two passing celestial bodies have minimal angular separation; (2) The moment when two solar-system bodies have the same celestial (ecliptic) longitude. A planet is said to be in conjunction with the Sun when it has the same longitude as the Sun as seen from the center of the Earth. When the planet is between the Earth and the Sun it is in inferior conjunction; when the Sun is between the planet and the Earth, it is in superior conjunction.

constellations.   The stars form interesting patterns which are well known to many people. There are dippers, crosses, and a variety of other easily identifiable figures. In the original sense the constellations are these configurations of stars. Two thousand years ago the Greeks recognized 48 constellations with which they associated the names and forms of heroes and animals of their mythology. The 48 original constellations, with certain changes made by the Greeks, are described in Ptolemy's *Almagest* (about A.D. 150) which specified the positions of stars in the imagined creatures. The original constellations did not cover the entire sky. Of the 1,028 stars listed by Ptolemy, 10 percent were "unformed" that is, not included within 48 figures. In various star maps that appeared after the beginning of the 17th century, new configurations were added to fill the vacant spaces, and they received names not associated with mythology. At present 88 constellations are recognized.

Constellations of Spring (March 21–June 21)

Coma	Leo
Bernices	Lynx
Corvus	Sextans
Crater	Ursa Major
Hydra	Ursa Minor

Constellations of Summer (June 21–September 23)

Aquila	Draco	Scutum
Bootes	Equuleus	Serpens Caput
Canes Venatici	Hercules	Serpens Cauda
Capricornus	Libra	Virgo
Corona	Lyra	Vulpecula
Borealis	Ophiuchus	
Cygnus	Sagittarius	
Delphinus	Scorpius	

Constellations of Fall (September 23–December 22)

Andromeda	Cetus
Aquarius	Lacerta
Aries	Pegasus
Cassiopeia	Pisces
Cepheus	Triangulum

Constellations of Winter (December 22–March 21)

Auriga	Gemini
Cameloparadalis	Lepus
Cancer	Monosceros
Canis Major	Orion
Canis Minor	Perseus
Eridanus	Taurus

constellation, zodiacal.   One of the twelve constellations through which the Sun and planets usually appear to move.

continuous spectrum.   A featureless spectrum consisting merely of a band of colors emitted by an incandescent solid, or a liquid or gas under high pressure.

continuum.   The continuous distribution of colors or wavelengths making up a more or less uniform background in a spectrum, as opposed to the discrete emission lines or absorption lines superimposed on such a background. The photons involved in ionization and recombination have a continuous distribution of energies and therefore of wavelengths; when detected, they thus form a continuum in the spectrum.

convection.   Transport of heat energy through the motion of the medium (gas or liquid). Occurs in the Earth's atmosphere when masses of warm air rise and masses of cold air fall to take their place. Less familiar examples include convection inside the Sun and Earth. Whether or not convection is effective depends on the rate of temperature change with altitude or depth. If this temperature change, or gradient, is large enough, convection will be the major process transporting energy from warmer to cooler regions.

coordinate.   A distance or angle characterizing the position or location of an object. Coordinates x and y are generally used on a plane surface; longitude and latitude on the Earth's surface; rich ascension and declination, or galactic longitude and latitude in the sky. Three coordinates are needed to specify a position in three-dimensional space.

Copernican System.   The theoretical model of the solar system in which the planets and the Earth revolve about the Sun, and the Earth rotates about its axis from West to East. Nicolaus Copernicus was a 16th century Polish mathematician and astronomer. He discovered that Aristarchus had suggested a universe with the Sun at the center—a heliocentric instead of a geocentric universe. Copernicus adopted this system which simplified the problem of explaining the motions of the planets by having them describe orbits around the Sun. The present heliocentric solar system is called the Coperinican system, although it was changed and improved about 100 years later by Johannes Kepler. (See Fig., p. 57.)

Copernicus, Nicolaus (1473–1543).   Nicolaus Copernicus inaugurated a new era in astronomy by discrediting the ancient theory of the central, motionless Earth. In his book *On The Revolutions of the Celestial Bodies*, published shortly before his death, he showed that all these motions could be interpreted more reasonably on the theory of the central Sun. He assured that the Earth revolves around the Sun once in a year and rotates daily on its axis.

Copernican revolution.   The assimilation during the Rennaissance of the idea that the Earth is not the center of the universe. This idea caused a profound reassessment of humanity's role in the cosmos.

coriolis correction.   A correction applied to an assumed position or trajectory to allow for apparent acceleration due to coriolis acceleration.

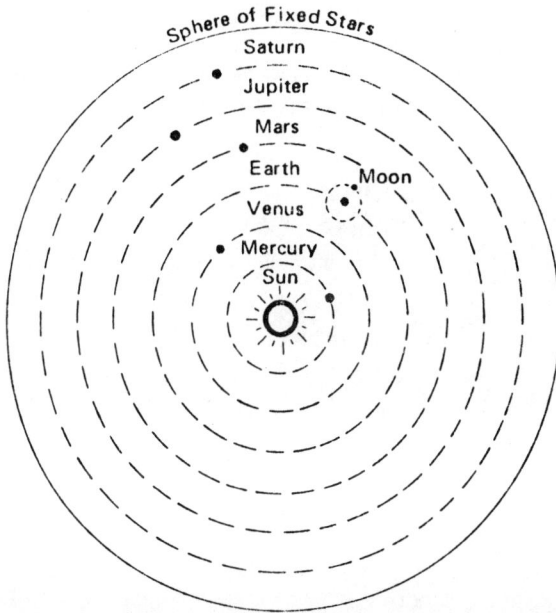

Sphere of Fixed Stars

coriolis force.   An imaginary force perceived by an observer in a rotating system observing a moving body or particles in the rotating system. It is associated with the movement of the masses involved, perpendicular to the axis of the primary rotating system, and is important in tracking weather systems on the Earth.

corona.   The outermost part of the Sun's atmosphere which becomes visible during a total eclipse by the Moon. It changes in shape periodically in sequence with the sunspot cycle and extends outwards up to thirty solar radii. It contains highly ionized gases incluing gaseous iron, nickel, and calcium at a temperature of about 1 to 2 million° K.

Corona Australis (The Southern Crown).   A small circlet of stars not as conspicuous as Corona Borealis but worthy of being ranked as a separate constellation, and sufficiently far north to be included in the original 48 groups of Ptolemy, who assigned 13 stars to it. In the small area of 128 sq deg is an ellipse of fairly bright stars which serve to mark the constellation in the Milky Way. No open star clusters occur, but there are two globular clusters. Much diffuse dark nebulosity is present also, lit in places by immersed stars. There are two planetary nebulae.

CORONA AUSTRALIS

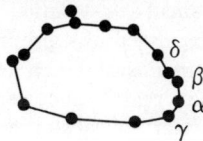

Corona Borealis (The Northern Crown). An effective little circlet of stars between Bootes and Hercules which has an area of 169 sq deg. It is one of the Ptolemaic northern groups to which 8 stars were assigned and was known to the early Greeks as the wreath. Only later was the qualifying adjective "northern" added, to distinguish it from its southern counterpart.

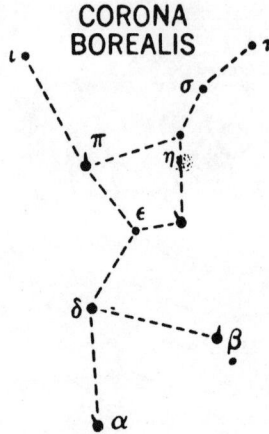

coronagraph. An instrument designed to allow photography of solar corona, which is almost a million times fainter than the disk of the Sun. It was invented in 1931 by the French astronomer B. Lyot. It consists essentially of two telescopes in tandem, one behind the other. The first telescope forms an image of the Sun artificially eclipsed by a metal disk at the focus. The second lens images the first lens on the third lens, which images the light from the Sun's atmosphere that extends beyond the edge of the eclipsing disk upon a photographic film or plate. Diaphragms remove scattered light and spurious faint images from the system.

corpuscles. Obsolete term for atomic and subatomic particles.

corpuscular theory of light. The hypothesis by Sir Isaac Newton that light consists of a stream of minute particles emitted by luminous bodies at very high velocities and that the sensation of light is due to the bombardment of the retina of the eye by these particles.

correcting lens (telescope). The lens placed at the front of a catadioptric telescope; used to correct spherical abberation of the primary mirror.

Corvus (The Crow). A small constellation of ancient origin to which Ptolemy assigned 7 stars. It is distinguished by 4 prominent stars in a trapezium. The group lies between Hydra and Virga; the modern area is 184 sq deg. The main telescopic interest of the constellation centers about some of the double stars, a planetary nebula NGC 4361 with a clear central star, and 4 distant galaxies. (See Fig. p. 59.)

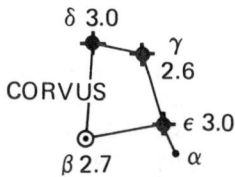

δ 3.0
  γ
  2.6
CORVUS
  ε 3.0
β 2.7   α

cosine law of illumination.   A geometric relationship between the flux F of radiation or particles striking a surface, the angle of incidence θ, and the illuminance I received by the surface. The relation is $I = F\ cos\ \theta$.

cosmic radiation (at radio wavelengths).   When it arises from thermal radiation, the cosmic-radiation spectrum is that of black-body radiation and is strongest in the centimeter lengths. When it arises from nonthermal radiation (the second type, which may be mainly the sort of radiation emitted by fast-moving particles revolving in the magnetic field of the synchroton in the radiation laboratory), it is much stronger in the meter wavelengths than in shorter ones and is separated thereby from the thermal radiation.

cosmic rays.   Extremely high-energy subatomic particles which bombard the Earth from all directions. Protons constitute about 85 percent of primary cosmic rays. About 14 percent are some nuclei of helium atoms (alpha particles) and about 1 percent, electrons. Other particles include a few nuclei of heavier atoms. Most cosmic rays strike and break up atomic nuclei in the upper atmosphere. Physicists measure and count cosmic rays with Geiger counters, cloud or bubble chambers, photographic emulsions, or other devices that can reveal the sudden production of free ions or electrons. Lower energy cosmic rays (10^7 to 10^{10} eV) come from the sun; higher energy ones (10^{10} to 10^{16} eV) come from elsewhere in the galaxy, possibly from supernovae.

cosmic-ray burst.   An extensive production of ionization from a cosmic ray striking an atom or molecule in the atmosphere.

cosmogony.   (1) The science of the origins of galaxies, stars, planets, and satellites. (2) The investigation of the origin of the whole physical universe viewed as a single system.

cosmologist.   A person who investigates the laws and structure of the physical universe.

cosmology.   The study of the universe and its arrangement as a whole. The problem of cosmology is to collect all the known data about the universe and its various parts in order to put together a self-consistent hypothesis about its past and future structure.

cosmonautics.   The science of space travel (transliteration of the Russian term equivalent to astronautics).

cosmos. The physical universe.

coudé (pronounced coo-da'y). Refers to one of several methods for using a reflecting telescope to obtain a long focal length and a fixed image. The image is formed in the coudé room under the floor of the telescope. The coudé focus is the place where an image is formed that remains fixed when the telescope is moved. The name is derived from the French word *coudé* meaning elbow.

couder screen. Device worked out by Andre Couder consisting of a perforated screen placed over a mirror. The perforations are placed at the points where particular shadow patterns are to be observed and are of different shapes to accommodate the various shapes of the shadows they expose.

counterglow. A very faint, roughly elliptical glow extending about 20° along the eliptic and 10° in celestial latitude. It is centered 2° or 3° west of the point in the sky opposite the Sun and is variable in form and composition. The glow is visible to the unaided eye under the most favorable conditions.

Crab Nebula. A nebula in our galaxy containing a neutron star pulsar at its center with surrounding gases streaming outwards at a speed of about 1000 km/sec. It has been identified as the debris of a supernova that was recorded in A.D. 1054 in detail by Chinese astronomers. It is one of the most prominent radio transmitters in the sky. The central pulsar NP 0532 pulses every 0.0331 seconds.

Crater (The Cup). The southern cross is the smallest of the constellations with an area of only 68 sq deg. The early Portuguese navigators saw in it the symbol of their faith, and the mystery of the unknown lent it an additional charm in the minds of those from whom the southern skies were hidden. Crux lies almost entirely in the Milky Way which is very bright in this region and so renders conspicuous the large irregular dark nebula immediately known as the Coal Sack. The constellation contains some beautiful stars, as well as 10 open clusters, including the vivid NGC 4755.

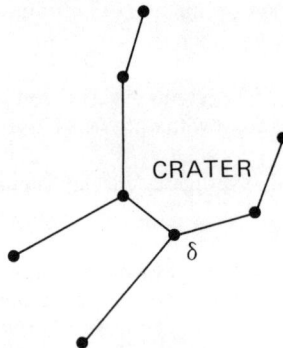

CRATER

craters. (1) Circular structures believed to be (mostly) scars of meteorite impacts on the Moon and planets. They have depressed floors, slightly raised rims, and are surrounded by ejected material. They range in size from microscopic (as found on lunar samples) to structures exceeding 1,000 km in outer diameter. The most famous craters, those detected by telescope on the Moon, range from a few km to 200 km across. Originally regarded as a curiosity of the Moon, they have now been found on every terrestrial planet and satellite, including the Earth, in various states of preservation or erosion. (2) A circular depression of volcanic origin usually found near the summit of a volcanic mountain or cinder cone. Confusion between volcanic craters and meteorite-impact craters marked the early theories about the craters of the Moon.

craterlet.   A small crater on the Moon's surface ranging in diameter from a few feet to 2 or 3 miles. The term originated with telescopic observers for the smallest (~1-mile) craters they could see; it is little-used today.

crescent.   The phase of the Moon or a planet when less than half the visible disk is illuminated by the Sun.

Crux (The Cross).   A splendid, compact constellation lying in that part of the Milky Way nearest the south celestial pole containing the famous dark nebula known as the "coal sack," a dust cloud blotting out the stars beyond it. Crux is a small constellation, but with its clearly cruciform shape it is unmistakable.

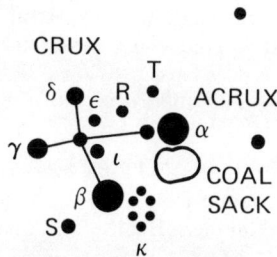

CRUX

T

δ ϵ R ACRUX α

γ ι

β COAL SACK

S

κ

culmination.   The moment when a celestial object crosses the meridian and has its greatest elevation above the horizon.

cumulonimbus clouds.   Described as heavy masses of clouds with great vertical development whose cumuliform summits rise in the form of mountains or towers. The upper parts have a fibrous texture and often spread out in the shape of an anvil.

cumulus clouds.   Literally, an "accumulation, a stack or heap," and is just that. Defined as "thick clouds with vertical development; the upper surface is dome-shaped and exhibits protuberances, while the base is perhaps the most familiar. The common forms of cumulus are easily identified. Like puffy balls of cotton that parade along in loose formation, they seek the same stratum or level and stretch out in line for miles. Cumulus clouds have many shapes and forms.

curvature of field (telescope).   A lack of uniform sharpness throughout the field of view. If objects at the center of the field are in sharp focus, those at the edge will be blurred and vice versa. The defect increases with the square of the distance from the center of the field.

curved path error.   The difference between the length of a ray refracted by the atmosphere and the straight-line distance between the ends of the rays.

curve of growth.   The relationship between the intensity of an absorption line (the amount of energy absorbed by it) and the total number of atoms acting to produce it. The curve of growth indicates how an absorption line will "grow" in intensity as the concentration of a given absorbing gas increases.

curvilinear coordinates.   Any linear coordinates which are not Cartesian coordinates. Examples of frequently used curvilinear coordinates are polar coordinates and cylindrical coordinates.

cusp.   The pointed "horn" or tip of a crescent moon or planet.

cusp cap.   A bright polar region on Venus, often reported by telescopic observers prior to spacecraft flights, and confirmed in ultraviolet spacecraft photos.

cyanometry.   The study and measurement of the blueness of the sky. The characteristic blue color of clear skies is due to preferential scattering of the short wavelength components of visible sunlight by air molecules. Presence of foreign particles in the atmosphere alters the scattering processes in such a way as to reduce the blueness, making the color a more neutral white. Hence spectral analysis of diffuse sky radiation provides useful information concerning the scattering particles.

cyclone.   An atmospheric system in which the wind blows spirally around and in toward a center.

cyclonic.   Having a sense of rotation about the local vertical the same as that of the Earth's rotation; that is (as viewed from above), counterclockwise in the Northern Hemisphere, clockwise in the Southern Hemisphere, undefined at the equator; the opposite of anticyclonic.

Cygnus (The Swan).   One of the larger of the old constellations to which 17 stars were assigned by Ptolemy. It lies in the Milky Way following Lyra in a rich region of the sky and may be distinguished by the large cross (known as the Northern Cross) made by its bright stars. The modern area is 804 sq deg. This constellation contains many fine pairs and although globular clusters are absent, at least 28 open star clusters have been distinguished.

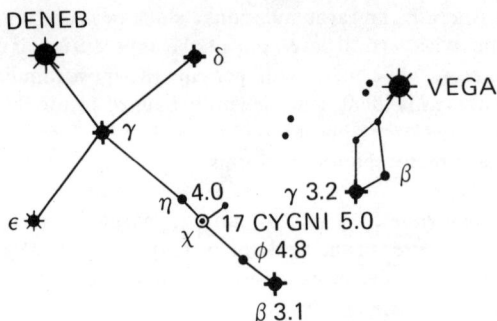

Cygnus A.   Third strongest celestial radio source (after the Sun and Cas A), associated with a remote galaxy with an unusual double nucleus.

Cygnus Loop Nebula.   NGC 6992, a wispy arc of filamentary gas ejected from a supernova about 20,000 years ago. It is estimated to be 2500 light-years away and is identical with x-ray source Cyg χ-5.

cylindrical projection.   A map projection produced by projecting the geographic meridians and parallels onto a cylinder which is tangent to the surface of a sphere and then developing the cylinder into a plane.

D

D-layer. An ionospheric layer starting at about 100 km and reflecting radio broadcasts.

D-lines. Prominent yellow spectral lines (5890 and 5896 A) arising from sodium.

Dall–Kirkman telescope. A variation of the Cassegrainian telescope. It makes use of a spherical secondary mirror and an oblate spheroid for a primary.

dark adaptation. Adjustment of the eye to darkness, essential for satisfactory observation of faint objects.

Dawes limit. The smallest angular separation (α'') of two stars in which each is still observable with a telescope of a given aperture (A cm). Daws limit is approximately $\alpha'' = 12/A$.

day. (1) A measure of time based upon one complete rotation of the Earth. A solar day is measured from a transit of the Sun across a given meridian to its next successive transit across the same meridian. The mean solar day of $24^h00^m00^s$ averages slight variations in the solar day, resulting mostly from the Earth's noncircular orbital motion. (2) For any planet, the rotation period.

day, Julian. The date expressed as the number of days since January 1, 4713 B.C. Julian days are used to keep chronological records of events such as eclipses over long historical intervals.

day, sidereal. Duration of Earth's rotation with respect to the stars, $23^h56^m4^s.091$.

declination. The angular distance of an object north or south of the celestial equator. It is measured in degrees and minutes. North declination is indicated by a + sign and the south declination by a − sign. Parallels of declination are the celestial equivalents of parallels of latitude on Earth.

declination axis. The axis about which a telescope is turned to make adjustments in declination. Most telescopes are mounted so that they can turn on two axes to follow the circles of the equatorial coordinate system—with one declination axis and one polar axis parallel to the Earth's axis.

declination circle (telescope). A setting circle divided into units of 0° and 90° to measure declination.

deferent. The hypothetical circular path along which the center of a planetary epicycle moved in the Ptolemaic model of the solar system.

definition (telescope).   Quality or sharpness of a telescope image over the whole field of view. Definition depends on the quality of the optics.

deflection of the vertical.   The angular difference at any place between the direction of a plumb line (the gravitational vertical) and the perpendicular (the normal) to the reference spheroid. This difference seldom exceeds 30 seconds of arc.

degenerate electrons.   Form of electrons in ultrahigh-density objects, such as white dwarf stars. Electrons can occupy energy states only in pairs, and once a pair of electrons is in a particular state, this state is closed to all others. So, as the density increases, electrons begin to occupy energy states beginning with the lowest available states. The distribution of electron energies is unlike that found in an ordinary gas.

degenerate gas.   A gas whose properties are determined by degenerate electrons. Instead of depending on density and temperature as in ordinary gas, the pressure in a degenerate gas depends to a first approximation only on the density, and the material behaves unlike an ordinary gas, liquid, or solid.

degree, angular.   A unit of angular measures. There are 360 degrees in a complete circle.

degrees, Celsius.   Temperatures are measured in degrees, and there are several scales in use. Most of the world uses degrees Celsius (or centigrade) defined by the melting or freezing point of water (0° C) and its boiling point (100° C). Absolute zero then is about −273° C. Fahrenheit degrees (F) can be derived from Celsius degrees (c) by: $F = 9/5\ c + 32$.

degrees, Kelvin.   Science bases its scale (degrees Kelvin) on the size of the centigrade unit but starts with 0 K at the absolute-zero point. Thus, water freezes at about 273 K and boils at 373 K. The abbreviation K is preferred over °K.

Deimos.   A satellite of Mars about $10 \times 12 \times 16$ km in diameter and 23,000 km from the planet's center.

delayed neutrons.   Neutrons emitted by excited nuclei in a radioactive process, so-called because they are emitted an appreciable time after the fission.

Delphinus (The Dolphin).   A small constellation lying between Aquila and Pegasus. It has been mapped since ancient times and Ptolemy assigned 10 stars to it. In spite of

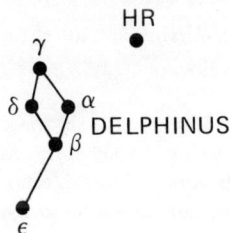

its small size, Delphinus contains some interesting double stars, two planetary neculae and two globular clusters.

Delta Cepheus.   A noted variable star being the type-example from which the Cepheid variables are named. These stars vary in brightness in short periods, and their fluctuations are due to periodic pulsations caused by internal instabilities. They are giant stars in both size and brightness. Delta Cepheus's variations were discovered in 1784 by English astronomer John Goodricke at age 19. It varies from magnitude 3.6 to 4.3 and back in a period of 5.4 days.

Delta Libra.   A variable star of the Algol type. During its period of two days and eight hours, its light varies from the 5th to 6th magnitudes, while it is being regularly eclipsed by a dark companion.

Delta Orionis.   An eclipsing binary with a range of from 2.20 to 2.35. The brightest component is a type O supergiant. This is much too slight to be noticeable with the naked eye under ordinary conditions, and it is surprising to find that its variability has been known for a long time. Sir John Herschel first pointed it out in 1834 and in Chamber's catalogue of 1890 the range was given as 2 1/4 to 2 3/4. The pair is about 1500 light-years away.

Deneb (Cygnus).   See Alpha Cygni.

Denebola.   See Beta Leonis.

density.   The mass of a substance per unit of volume; for example, the number of grams per cubic centimeter. A dense substance is one in which a large quantity of matter occupies a small volume of space. By definition, water has a density of 1.0 grams/cm^3.

depressed pole.   The celestial pole below the horizon.

descending node.   Point at which a planet, asteroid, or comet crosses from the north to the south side of the ecliptic plane, or a satellite crosses to the south side of the equatorial plane of its primary.

dewcap.   A protecting collar which extends beyond the outer end of a refracting telescope to prevent dew from condensing on the surface of the lens.

diagonal mirror.   A small, flat mirror used to reflect the main cone of light into the eyepiece of a Newtonian telescope.

diaphragming.   A technique to improve the sharpness of a telescope image by covering the telescope aperture with a ring-shaped plate or diaphragm, thus reducing the effective aperture. While this technique reduces effects of bad seeing or aberrations in an imperfect optical system, it also cuts down the brightness of the image.

dichotomy.   A configuration when an object is observed at phase angle 90°. The object is half illuminated and its terminator is straight.

differential correction.   In celestial mechanics, a method for finding, from observed residuals minus the computed residuals (O–C), small corrections which when applied to the orbital elements or constants will reduce the deviations from the observed motion to a minimum.

differential rotation.   Motion where certain parts of a system (such as outer regions or regions at high latitudes) rotate at a velocity different from other regions (nearer the center, or lower the latitude).

differentiation.   Any chemical or geologic process that modifies composition of a material from its parent state by separating different chemical constitutents. Examples are loss of hydrogen and volatiles from primitive planetary material, and separation of different minerals in magma during volcanic processes.

diffraction.   The bending of light around an obstruction of any size, such as the edge of the Moon, the edge of a telescope aperture, or the edges of small particles. Diffraction is one evidence of the wave properties of light. Diffraction prevents a telescope from producing a perfect point image even from a perfect point source of light.

diffraction grating.   A system of parallel edges designed to produce diffraction. Usually it is a piece of flat glass with close parallel lines having separations comparable to the wavelength of light scribed on its surface. Since the amount of diffraction is different for different wavelengths, reflection or transmission of light by the grating produces a spectrum. Thus the grating can be used as a prism.

diffraction rings.   The rings that surround the image of a star or other point source of light seen in a telescope due to diffraction within the telescope.

diffuse nebula.   A reflection or emission nebula caused by a concentration of interstellar matter near a bright star.

diffuse sky radiation.   Solar radiation reaching the Earth's surface after having been scattered from the direct solar beam by molecules or particles in the atmosphere.

Dione.   A satellite of Saturn orbiting at a mean distance of 378,000 km. Its diameter is estimated to be 900 km.

dip angle.   (1) The vertical angle between the true horizon and the apparent horizon. (2) The angle between a geologic surface, such as a fault or stratum, and the horizontal.

dipole.   A system composed of charges separated by some distance (i.e., a bar magnet).

direct motion.   Eastward or counterclockwise motion of a planet or other object as seen from the North Pole (motion in the direction of increasing right ascension).

direct solar radiation.   The portion of the radiant energy received by an instrument direct from the Sun, as distinguished from diffuse sky radiation, terrestrial radiation, or radiation from any other source.

discrete radio source.   A source of cosmic radio waves of small angular extent.

dish.   A reflector (for reflecting or collecting radio waves), the surface of which is part of a sphere or paraboloid; a parabolic reflector type of radio or radar antenna. Its surface may consist of wire mesh which reflects radio waves.

dispersion.   (1) The separation of electromagnetic radiation by wavelength due to a characteristic of the medium. (2) Frequency-dependent retardation of radio waves as they pass through ionized gas.

dissociation.   The separation of a complex molecule into its constituent atoms, molecules, or ions by collision with a second body or by absorption of a photon.

dissociative recombination.   Recapture of an electron by a positive molecule resulting in splitting of the molecule into two neutral particles.

distance modulus.   Difference between apparent magnitude m and absolute magnitude M, a measure of the distance of an object based on the assumption that space contains no absorbing dust or gas so that brightness is proportional to 1/distance 2. If there is no absorption, $m - M = 5 \log (r/10)$ where r is in parsecs.

distortion.   An optical defect which causes a straight line passing through the center of a telescopic field to become curved as it approaches the edges. The curvature can be convex around the center of the field (barrel distortion) or concave (cushion distortion), and becomes more pronounced as the distance from the center increases. Distortion is caused by unequal magnification of different parts of the image and is very similar to that seen at the corners of "wraparound" windshields in automobiles. It is not a serious defect when looking at stars, but can be intolerable in a telescope used for terrestrial observation.

diurnal.   Having a period of a day related to Earth's daily rotation. Daily.

diurnal aberration.   Aberration caused by the rotation of the Earth. The value of diurnal aberration varies with the latitude of the observer and ranges from zero at the poles to 0.31 seconds of arc at the equator.

diurnal circle.   The apparent daily path of a celestial body; a parallel of declination (for a fixed object).

diurnal motion. The apparent daily motion of a celestial body as observed from the rotating Earth.

diurnal parallax. The apparent change in direction of an object due to the shift in the position of an observer by the daily rotation of the Earth.

Doppler effect. A change in the wavelength and frequency of sound, light, or other wave radiation due to relative motion between the emitting source and the observer. The change in light wavelength $\Delta\lambda$, is a function of the relative velocity v, wavelength λ, and the speed of light c. For velocities much less than the speed of light it is given by $\Delta\lambda = \lambda v/c$. The effect is extremely useful in measuring many stellar properties.

Doppler–Fizeau effect. The Doppler effect applied to a source of light.

Doppler shift. A change in wavelength due to the Doppler effect.

Dorado (The Swordfish). Introduced as a constellation by Bayer in 1604 and contains the interesting naked-eye galaxy called Nubecula Major or the Greater Magellanic Cloud, at a distance of 150,000 light-years from the Earth. It also contains one of the most luminous known stars, S. Doradus, which is estimated to be almost 400,000 times more luminous than the Sun. It is a supergiant eclipsing binary in the Large Magellanic Cloud.

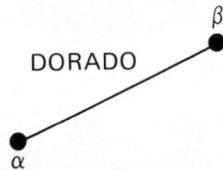

dosimeter. An instrument for measuring the ultraviolet in solar and sky radiation.

double stars. There are two kinds of stars that may be called double. In one type, two stars appear to be very close together, yet one may be a great distance beyond the other. Their apparent closeness is due to our line of sight. They are usually called visual double stars. The second type is a system of two stars moving together in space and revolving about each other. They are usually called physical binaries or just binaries. Many of both types present interesting telescopic views. Examples are listed below.

Star	Magnitude	Color
Zeta ζ Ursa Majoris	2.4 and 4.0	Mizar and Alcor, blue and blue (visual double)
Iota ι Cancer	4.2 and 6.6	Yellow and blue
Nu ν Draconis	5.0 and 5.0	Blue and blue
Epsilon ε Lyrae	4.7 and 4.5	blue and blue (visual)
Beta β Cygni	3.2 and 5.4	yellow and blue (Albireo)
Nu ν Scorpii	4.3 and 6.5	white and blue

Star	Magnitude	Color
Gamma γ Andromedae	2.3 and 5.1	yellow and blue
Beta β Capricorni	3.2 and 6.2	orange and blue

doublet. A close pair of absorption or emission lines in a spectrum.

"doughnut" shadow (telescope). The apparent ringlike characteristic of the oblate spheroid, paraboloid or hyperboloid which appears under the Foucault test.

Draco (The Dragon). The large constellation of the dragon is marked by a number of widely scattered bright stars and lies almost wholly north of 50° N declination. It is of ancient origin and Ptolemy assigned 31 stars to it. The modern area is 1,083 sq deg. There are many interesting pairs, but neither galactic nor globular clusters occur and there is no diffuse nebulae. The single planetary nebulae NGC 6543 is situated close to the northern pole of the ecliptic.

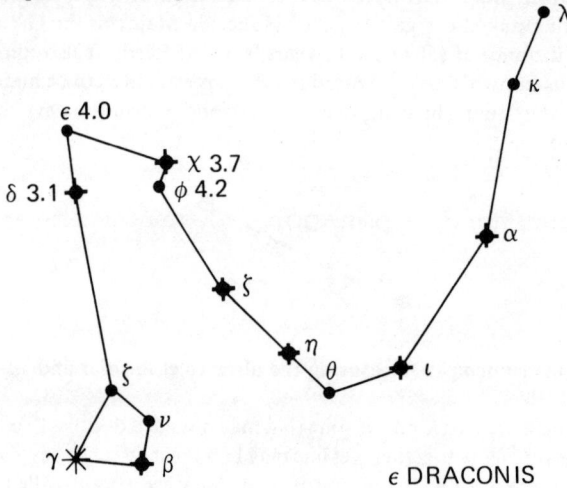

ε DRACONIS

Draconids. June Draconids are a meteor shower with maximum activity near June 28, associated with Pons–Winnecke comet. October Draconids were meteor showers in 1926, 1933 and 1946 when the Earth passed through the orbit of Giacobini–Zinner comet shortly before or after the comet itself.

draconic month. The average period of revolution of the Moon about the Earth with respect to the Moon's ascending node, a period of 27 days, 5 hours, 5 minutes, 35.8 seconds, or approximately 27 1/4 days. Also called nodical month.

draw tube (telescope). A freely moving tube that holds the eyepiece in a telescope.

dust. (1) In meteor terminology, finely divided solid matter having particles sizes generally smaller than micrometeorites, such as meteoric dust. (2) Interstellar grains, generally of micrometer dimension.

dust devil.   A column of swirling dust created when hot sun heats a dusty surface causing heated air to rise by convection. The masses of air swirl as they rise, picking up dust and swirling it high into the air. This phenomenon, common in terrestrial deserts, is believed to be involved in initiating dust storms on Mars.

dust storm.   On Mars, atmospheric phenomena in which some or all of the surface markings may be obscured by featureless clouds or airborne dust. They are commonly initiated as Mars swings through perihelion and the sun heats summer afternoon surfaces most strongly, possibly initiating dust devils or strong winds.

dwarf.   (1) A type of main-sequence star (a somewhat confusing usage, since white dwarfs are much fainter than main-sequence stars of similar spectral type). (2) A star less luminous than a main-sequence star of the same spectral type, usually called a subdwarf.

dwarf galaxy.   A galaxy of low luminosity.

dwarf nova.   A variable star, typified by U Geminorum, subject to nova-like outbursts at unpredictable intervals, usually between 16 and 300 days. These are believed to be close binaries in which mass is ejected from a giant onto a white dwarf.

dynamic height.   The height of a point in the atmosphere expressed in a unit proportional to the geopotential at that point. Since the geopotential at altitude z is numerically equal to the work done when a particle of unit mass is lifted from sea level up to this height, the dimensions of dynamic height are those of potential energy per unit mass.

dynamic parallax.   A value for the parallax (i.e., distance) of a binary star computed from the observations of the period and the angular dimensions of the orbit by applying the laws of motion.

dynamical mean sun.   A fictitious sun conceived to move eastward along the ecliptic at the average rate of the apparent Sun.

dynamo theory.   (1) The hypothesis first proposed by Balfour Stewart that the regular daily variations in the Earth's magnetic field result from electrical currents in the lower ionsphere, generally by tidal motions of ionized air across the Earth's magnetic field. (2) The hypothesis that the planets' magnetic fields derive from electrical currents produced by fluid motions in the planetary cores.

dyne.   A very small unit of force equaling the force necessary to speed up 1 gram by 1 cm/sec in 1 sec. Expressed as weight, it is about 1/1000 the weight of a gram on the Earth's surface.

dyne–centimeter.   A unit of energy equaling an erg. The energy expended by exerting one dyne of force to move an object a distance of one centimeter.

E

e. The base of the natural or Naperian logarithm system equaling 2.7183.

e-folding time. The time required during a phenomenon when a designated factor or component changes by a factor $e = 2.7183$.

early-type star. A hot star, usually implying spectral class O, B, A, or early F. The term originated with an early, incorrect theory of stellar evolution, but persists in current usage.

Earth. The Earth is a globe about 12,756 km in equatorial diameter, slightly flattened at the poles and bulged at the equator. Its irregular surface is 70 percent covered with water, and it is enveloped by an atmosphere mostly compressed within a height of 100 km above the surface. The Earth's mass is 5.98×10^{27} grams; its mean density is 5.517 g/cm^3. Aside from its atmosphere and hydrosphere, the Earth consists mainly of three parts: (1) the outer granitic and basaltic crust 4 to 35 km deep, (2) the mantle of denser rock extending 2,900 km below the crust, and (3) the core, believed to be mostly iron and nickel. The Earth's surface differs from other planets' surfaces due to two major factors: the presence of liquid water and the active geology driven by internal energy sources. Because water is present and the temperature permits it to be liquid, oceans exist and running water erodes land forms. Because of sluggish currents of material in the mantle, surface land forms are disrupted by movements known as plate tectonics, causing mountain formation in some regions and fracturing or subsidence in others. These forces of erosion and tectonics have obliterated most of the ancient meteorite-impact craters and lava-covered plains found on most other planets. Most of Earth's surface features are only hundreds of millions of years old, as opposed to the billions of years suspected for other planetary surfaces. Other properties of Earth include:

rotation period	$23^h56^m4.1^s$
obliquity (1973)	$23°26'34''$
gravity	$980 \ cm/5^2$
escape velocity at surface	11.19 km/s
albedo, oceans and land	≈ 0.2
albedo, clouds, snow	≈ 0.8
surface air pressure	$1.01 \times 10^6 \ dyn/cm^2$
age	4.6×10^9 yr.
core temperature	$\approx 6400°$ K

earth-crossing asteroids. See Asteroids, Apollo.

earth-grazers. Asteroids that pass very near Earth. Eros, with diameter about 35 km, is a famous example, approaching within about 23 million km. Hermes, with a diameter of about 1 km, passed the Earth in 1937 at a distance of about 600,000 km, less than twice the distance of the Moon. Icarus, about 1 km across, passed 6 million km away in 1968.

earthlight.   The illumination of the dark part of the Moon's disk by sunlight reflected onto the Moon from the Earth. Spectroscopic observations reveal that earthlight is relatively richer in blue light than is direct sunlight. The Earth's bluish color results from the fact that much of the total Earth reflection is backward-scattered light which according to Rayleigh's Law is relatively rich in blue and poor in red.

earthshine.   Earthlight.

eccentric.   Off center.

eccentricity.   A number which defines the shape of an ellipse. It is the ratio of the distance from center to focus to the semimajor axis. The orbit of Earth is nearly circular with an eccentricity of only 0.017.

eclipse.   Obscuring of one celestial body by another, either by direct superposition or by the casting of a shadow. Solar eclipses are of three kinds: total (when the Moon completely obscures the Sun), annular (when the Moon is too far away to cover the entire Sun and leaves a complete ring of sunlight visible around the edge of the Sun), and partial (when the Moon is "off center" and obscures only a portion of the solar disk). In all these cases, the Moon's shadow lies on the Earth. Lunar eclipses may be total (Moon fully inside Earth's shadow) or partial (Moon only partly covered by Earth's shadow). Eclipses also occur in other satellite systems. Viewers located inside an umbral shadow see a total eclipse; viewers in a penumbral shadow see a partial eclipse. See umbra and penumbra.

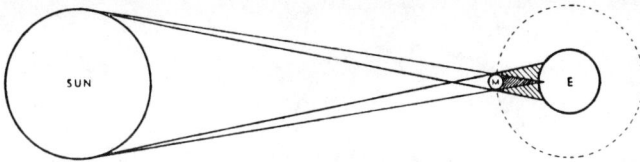

Eclipse of the Sun

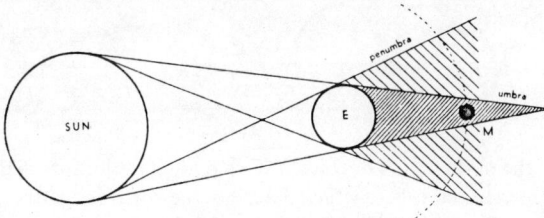

Eclipse of the Moon

eclipse seasons.   Intervals of time around the moments when the Sun passes one of the nodes of the Moon's path (i.e., when the Sun, Moon, and Earth are most nearly aligned in a straight line). Eclipses can occur only during those "seasons." Because the nodes regress westward the eclipse seasons come more than a half-month earlier from year to year.

eclipse year.   The length of the eclipse year is 346.62 days—the interval between two successive eclipse seasons or returns of the Sun to the same node.

eclipsing binaries.   A pair of stars moving around each other in orbits so oriented with respect to the observer that one star periodically eclipses (or more properly, occults) its companion. One reason for interest in eclipsing binaries is that they allow us to determine stellar radii directly. Depending on the inclination of the orbit we may have partial or total eclipses. Below are examples showing some of the varieties possible in light curves of eclipsing binaries. Eclipsing binaries generally appear in the telescope as single stars that become fainter while the eclipses are in progress. The periods in which known examples revolve and fluctuate in brightness average 2 or 3 days; they range from 17 1/2 minutes for the contact binary A M Canum Venaticorum to 27 years in the very exceptional case of Epsilon Aurigae.

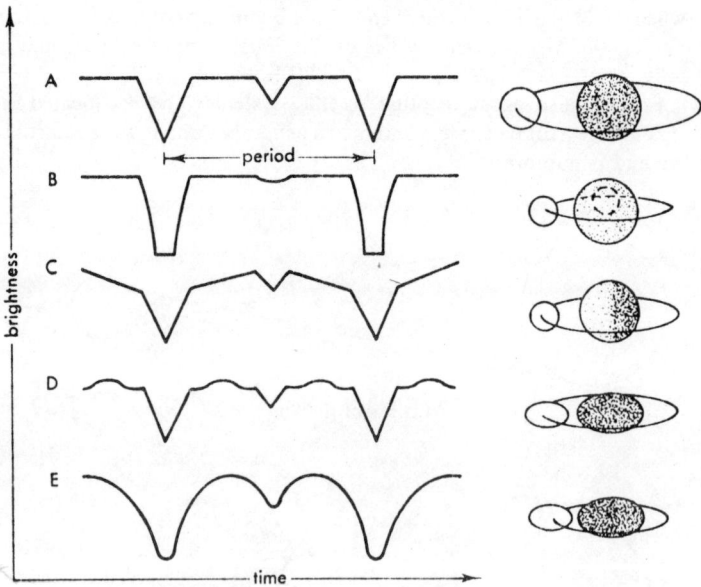

ecliptic.   (1) The plane of the Earth's orbit. (2) This plane's projection on the sky (i.e., the great circle on the celestial sphere which describes the apparent path of the Sun in the course of the year). The plane of the ecliptic is inclined to the plane of the equator at an angle of 23° 27'8".26. The angle is called the obliquity of the ecliptic. The constellations through which the ecliptic passes are called the signs of the zodiac.

ecliptic limit, lunar. The greatest angular distance of the Sun from the lunar node at which a lunar eclipse is possible, ranging from about 12° 15′ to 9° 30′ for umbral lunar eclipses.

ecliptic limit, solar. The greatest angular distance of the Sun from the lunar node at which it is grazed by the Moon, as seen from some point on Earth causing a solar eclipse. The value of the limiting angular distance varies with the changing linear distance from us and therefore the apparent sizes of the Sun and Moon, and with fluctuations in the angle between the Moon's path and the ecliptic.

ecliptic pole. On the celestial sphere, either of the two points 90° from the ecliptic. The north ecliptic pole is that lying on the north side of the celestial equator.

ecliptic system of coordinates. A set of celestial coordinates based on the ecliptic as the primary great circle. The poles are the north and south ecliptic poles. The angular distance north or south of the ecliptic, analogous to latitude, is the celestial latitude. Celestial longitude is measured eastward along the ecliptic from the vernal equinox through 360°.

ecosphere. A volume of space surrounding a star in which life might theoretically evolve, usually defined by a temperature region in which liquid water can exist. In the solar system it extends from near Venus' orbit to near Mars' orbit.

effective focal ratio (telescope). The ratio between the effective focal length of a compound system and the aperture of the mirror.

effective temperature (of a star). The temperature of a perfect radiator which has the same output of radiation per unit area as the star. The concept is useful because it gives a representative temperature of the surface layers of the star. In reality, the star's radiation comes from different layers with a range of temperatures.

eigenvalues. In quantum mechanics, specific values of certain parameters (such as energy) that are permitted, according to the equations and initial conditions assumed in the theoretical models. These parameters are said to be quantized; other values are not permitted. From the German "eigen," or characteristic.

E-layer. A division of the ionosphere, usually found at an altitude around 100 to 120 km. In this region the temperature rises with increasing altitude and short-wave radio waves are reflected. It is more pronounced by day than by night.

El Nath. See Beta Tauri.

electromagnetic radiation. Energy propagated through space or through material media in the form of an advancing disturbance in electric and magnetic fields existing in space or in the media. The terms "radiation" and "electromagnetic energy" refer to this type of energy. The term includes energy propagated at all wavelengths in the electromagnetic spectrum (q.v.).

electromagnetic spectrum.   The ordered array of known electromagnetic radiations extending from the shortest gamma rays, x-rays, ultraviolet radiation, visible radiation, infrared radiation, and including microwave and all other wavelengths of radio energy.

electrons.   The small, negatively charged particles in the outer region of any atom in motion about the nucleus. Mass = 9.11×10^{-28}; diameter = 5.63×10^{-13} cm; charge = 4.80×10^{-10} ESU. Sometimes electrons are removed from atoms and move independently, as in a stream forming an electric current, or as in ionization of a gas.

electron volts.   A unit of energy equaling 1.60×10^{-12} ergs used to describe atomic and subatomic particles. It is equivalent to the energy an electron acquires when accelerated across a difference of one volt in a vacuum. At this energy, an electron has a velocity of about 580 km/sec.

element.   (1) A chemically pure substance consisting of only one type of atom. Each element is defined by the number of protons in its atoms' nuclei. (2) One of the properties defining an orbit, such as semimajor axis.

ellipse.   An oval curve defined as a conic section formed by the intersection of a plane with a cone, the plane being inclined to the axis of the cone and cutting the sides of the cone. All curves defined by orbital motion of one body around another are ellipses.

elliptical galaxies.   The most common type of galaxy. They are elliptical in shape, lack spiral arms, and resemble in some properties the central bulges of spiral galaxies. They have many red giants, no young blue luminous stars, and little interstellar gas or dust. They include some of the most and least massive galaxies. Elliptical galaxies are classified observationally according to their apparent ellipticity (the ratio of their longer and shorter axes).

elongation.   (1) The angular distance of a body of the solar system from the Sun as seen from the Earth. The term is used primarily in connection with inferior planets. When the elongation is zero the planet is said to be in conjunction. (2) The moment when an inferior planet reaches its greatest angular distance from the Sun as seen from Earth.

Eltanin.   See Gamma Draconis.

emission.   The radiation of electromagnetic radiation generally involving a change in electrons from higher energy to lower energy. The excess energy is carried off by a photon whose wavelength is determined by the energy difference.

emission line.   A sharp excess of one color (wavelength) relative to adjacent colors in a spectrum. Emission lines are associated with hot or excited gas seen against cool or dark backgrounds, as in a flame or an emission nebula.

emission nebulae.   Huge masses of gas that absorb ultraviolet radiation from nearby hot stars and reradiate it in the form of emission lines, often by metastable or "forbidden" transitions. The largest emission nebulosities are almost always associated with O and B stars and in most cases with dense groups of extremely hot, luminous, newly-formed stars.

emission spectrum.   The spectrum normally produced by an incandescent gas at low pressure. It consists of bright emission lines against a dark background.

emission stars.   Stars having spectra in which bright emission lines of hydrogen and sometimes of other elements are present and frequently superposed in corresponding broader dark lines. About 4,000 such stars are known. About 10 percent of B0 and B3 stars and a progressively smaller proportion of B5 to A4 stars show emission lines in their spectra. In some cases the bright lines contain red in the spectra of what seems to be once-normal B stars that have subsequently disappeared.

emulsion.   The light-sensitive coating on film or glass plates, which is developed to show a negative image of the light falling on the film or plate.

Enceladus.   A satellite of Saturn orbiting at a mean distance of 238,000 km. Estimated diameter is roughly 560 km.

encounter.   A close, chance passing between two moving bodies (such as an asteroid and a planet) resulting in a gravitational perturbation of the motion.

enhanced radiation.   Increased radio wave or thermal radiation from the Sun, of several hours or days duration.

entropy.   A measurement of the amount of disorder in a system. As natural systems evolve toward equilibrium, entropy tends to increase, as predicted quantitatively by the second law of thermodynamics.

ephemeris.   (1) Catalogue of predicted positions of the planetary bodies, (2) *The American Ephemeris and Nautical Almanac*, which is such a catalog issued annually by the U.S. Naval Observatory.

Ephemeris Time.   A definition of time based on idealized motions of the Sun and Moon that was adopted and introduced in the astronomical ephemerides on January 1, 1960, to free astronomical computations from the effect of the irregularities of the Earth's rotation. One second of Ephemeris Time is a specified fraction of the tropical year 1900 (equals 31,556,925,9747 seconds of Ephemeris Time); the difference between Ephemeris Time and Greenwich Standard Time (Universal Time) is derived for each year from a comparison of the observed and computed position of the Moon.

ephemeris day.   86,400 ephemeris seconds.

ephemeris second. The fundamental unit of time of the International System of Units of 1960: 1/31556925.9747 of the tropical year defined by the mean motion of the Sun in longitude at the epoch 1900 January, 0 days, 12 hours.

epicycle. In the Ptolemaic system of the universe, the comparatively small circular path of a planet, the center of which moved on the circumference of a larger circle centered on or near Earth, called the "deferent." Such combinations of circles of ever-increasing complexity were used to represent and predict planetary motions until it was realized that planets moved in ellipses around the Sun.

epoch. In astronomy, an arbitrarily chosen moment in time to which measurements of position are referred. The epochs commonly used for star maps are 1855, 1900, 1920 and 1950. Corrections are necessary because the right ascension and declination vary continuously due to their dependence upon the position of the celestial equator. Also, the date for which an astronomical chart or catalogue has been calculated.

Epsilon Aurigae. An eclipsing binary with a period of about 27 years and an eclipse lasting 714 days. It is an extraordinary system made up of a highly luminous supergiant (60,000 times solar luminosity) and an immense tenuous companion about 35 A.U. away. The companion cannot be seen, but is thought to be a star surrounded by a 20 A.U.-radius, disk-shaped, dusty nebula in which planetary bodies may be forming. The large size of the nebula explains the great length of the ellipse.

Epsilon Bootes. Better known as Mirac, is a double star of magnitude 2.4 and 5.1, and separation about 3."6. Its components of orange and pale green are so pleasing a sight as to earn for it the title of Pulcherrima ("very beautiful"), but a small telescope is necessary to separate them. The star is sometimes called Izar and is about 110 light-years away.

Epsilon Cygnus. Epsilon was named Gienar, "the Wing." It marks the right wing of the Swan, as Delta marks the left wing. Its visual magnitude is 2.5 and its distance is about 75 light-years. It is a K giant.

Epsilon Lyra. The famous "double–double" star in Lyra. If a visual observer looks carefully he may be able to see that it is two stars about 208 seconds apart, each of about magnitude 14 1/2. A telescope shows that each of the two is a double star, one with separation 3 seconds (magnitude 5.0 and 6.0) and the other with separation 2 seconds (magnitude 5.1 and 5.4). The object is considered a good test of seeing and small-telescope optics.

Epsilon Pegasi. The star Enif, of visual magnitude 2.4. In the last century, Schwabe claimed that it varied between magnitude 2 and 2 1/2 in a period of 25 3/4 days, but this is not confirmed. It is a K supergiant about 800 light-years away.

Epsilon Sagittarius.   Kaus Australis, the southern and brightest of the 3 stars making the archer's bow. It is of visual magnitude 1.8 and spectral type B9. Its distance is estimated to be 160 light-years.

Epsilon Scorpio.   The second star below Antares is Epsilon, which has a visual magnitude 2.3. It is a double, the components being of magnitude 4.5 and 7.5 and white-gray in color. A 5th magnitude star lies very close to the largest component, making it an apparent triple star, the light from the three so blending as to appear to the naked eye as a single star.

Epsilon Ursa Majoris.   See Alioth.

equant.   A point on the plane of an orbit about which an epicycle or body revolves with uniform angular velocity (in the outmoded Ptolemaic theory of planetary motion).

equation.   (1) In astronomy, a small correction to observed values to remove the effects of systematic errors in an observation. (2) Any statement of equality between two quantities.

Equation of Time.   The interval by which the true Sun is ahead or behind the mean Sun. The Sun's apparent motion in the sky varies throughout the year because the Earth's speed in its elliptical orbit varies slightly from perihelion to aphelion, while the mean Sun is assumed (by definition) to travel with a constant speed equal to the average speed of the true Sun. The interval never exceeds 17 minutes.

equations of motion.   A set of equations which give information regarding the motion of a body or of a point in space as a function of time when initial position and initial velocity are known.

equator.   The line joining all points on a sphere's surface equidistant from the two poles of the axis of rotation. The centrifugal force and angular velocity are greatest at the equator.

equator, celestial.   A great circle on the celestial sphere midway between the two apparent poles of rotation. It corresponds to the intersection between the plane of the Earth's equator and the imaginary celestial sphere.

equatorial bulge.   The excess of a planet's (or star's) equatorial diameter over the polar diameter caused by deformation of the object produced by its spin.

equatorial mounting (telescope).   An equatorial mounting has two axes around which the telescope may move. The polar axis points toward the celestial poles parallel to the Earth's axis of rotation. A clock or motor slowly rotates this axis to turn the tele-

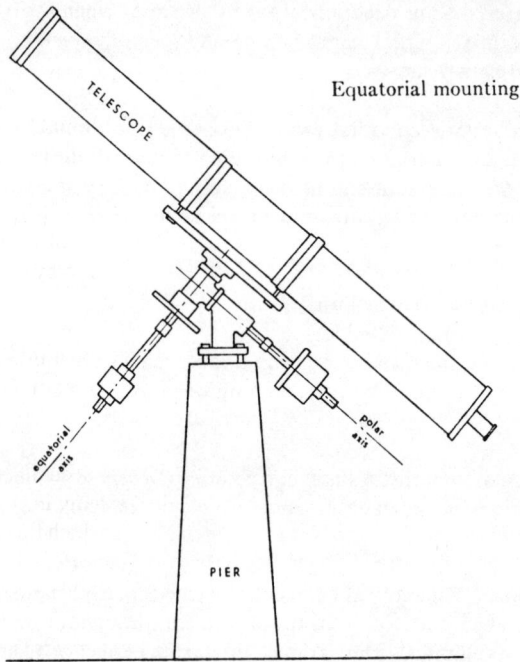

Equatorial mounting

scope westward, at the exact rate necessary to compensate for the Earth's rotation eastward. The result is that the telescope, always pointing toward the same star, is fixed with respect to the celestial sphere, while the Earth is turning under it.

equatorial satellite. A satellite whose orbit plane coincides (or almost coincides) with its planet's equatorial plane.

equatorial system. A set of celestial coordinates based on the celestial equator as the primary great circle (i.e., declination and right ascension).

equilibrium spheroid. The shape that the Earth would attain if it were entirely covered by a tideless ocean of constant depth which would flow until it achieved a surface shape with no further net deformational forces.

equinoctial. Pertaining to the equinoxes; east–west.

equinoctial colure. The great circle of the celestial sphere through the celestial poles and equinoxes; the hour circle of the vernal equinox.

equinoctial hours. Hours of equal length, whether reckoned by day or by night.

equinoctial points. The First Point of Aries and the First Point of Libra. One of the two points of intersection of the ecliptic and the celestial equator.

equinoctial time.   The time reckoned from the moment when the mean sun is at the mean vernal equinox.

equinox.   The points on the celestial sphere where the Sun's center crosses the equator; vernal from south to north (about March 21), autumnal from north to south (about September 23). Therefore, those points where the ecliptic intersects the celestial equator.

equivalent focal length (telescope).   The useful focal length of a compound optical system. (Obtained by multiplying the focal length of an eyepiece by the magnification it provides when used with the given telescope).

equivalent width.   The width of an absorption line with zero intensity and rectangular profile, whose area (total absorption) is equivalent to that of the true absorption line.

Equuleus (The Little Horse).   A constellation said to have been made by Hipparchus about 150 B.C. from some small stars near Delphinus. Ptolemy in the *Almagest* three centuries later assigned only four stars to it, the brightest of which form a small, narrow triangle which is not conspicuous. This smallest of the northern constellations with 72 sq deg is only slightly larger than the southern Crux.

Erfle (telescopic eyepiece).   This eyepiece consists of six elements and gives very good definition and wide field when it is made well. There is a slight loss of light due to the large number of lenses. It is one of the best wide-angle eyepieces.

erg.   The unit of energy cgs metric system; the work done in pushing 1 gram so that it speeds up from rest to 1.414 cm/sec, or in lifting 1/1000 gram 1 centimeter. It has been likened to the energy of a flea colliding with a wall.

Eridanus.   An ancient constellation. Ptolemy assigned 38 stars to it and the modern area is 1,138 sq degrees (among the largest of the star groups). Eridanus is rich in fine double stars but contains neither open nor globular clusters. There are no diffuse nebulae; many extra galactic nebulae occur but almost all of these are small and faint. (See Fig., p. 82.)

Eros.   An earth-approaching asteroid known since 1898. Eros has an orbit with semimajor axis 1.5 A.U. and period 1.8 years. Its orbit is eccentric enough so that a favorable opposition it can approach the Earth within about 22,000,000 km.

eruptive variable.   A variable star in which the irregular observed light variations are produced by some eruptive or explosive process within the star (i.e., supernovae).

escape velocity.   The least velocity a particle must have at a distance from some body to leave it (escape) and never return. For Earth's surface or near-earth orbit: 11.2 km/sec.

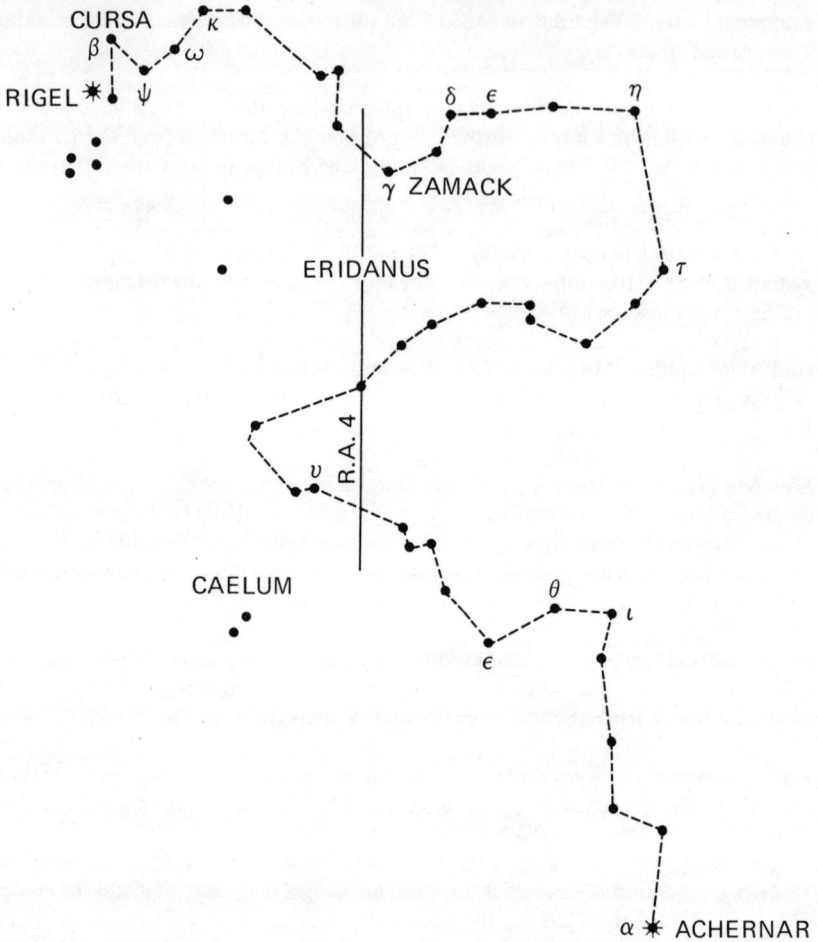

Eta Aquarids (Aquarius).   A meteor shower radiating from a point between Alpha and Eta Aquarii. It is visible for a short time before sunrise between April 29 and May 2.

Eta Aquila.   Eight degrees below Altair is Eta, an interesting short-period variable star which ranges from magnitude 3.5 to 4.7 and nearly doubles in brightness in the course of a single week; these changes take place with great regularity. It is a Cepheid variable.

Eta Carinae (variable star).   The most famous "blaze star" lying in the southern hemisphere. Eta Carinae was first catalogued by Edmund Halley, who noted it is a 4th magnitude object while observing it from the island of St. Helena in 1677. The Greek astronomer Ptolemy, who drew up a star catalogue in the second century A.D., had not noted a star in its place. Drastic changes soon followed. In 1687 and again in 1751 it was seen as a 2nd magnitude object, fading down between times to its more usual luster; in 1827 it shot up to the 1st magnitude, declined slightly, and finally summoned a burst of energy that brought it up almost to the level of Sirius, the brightest star in the sky. After holding this position for several years, this extraordinary star began to fade until in 1868 it was lost from naked-eye view. By this time it had reached a minimum magnitude of 7.6 at which it has remained to the present time. Telescopes reveal a great chaotic nebula around the star, apparently consisting of gas thrown off in explosive eruptions.

ether.   The medium once believed to be required to transmit electromagnetic radiation through space. Experiments by Michelson and Morley in 1887 showed that such a fixed medium does not exist.

Eulerian angles.   A system of three angles which define with reference to one coordinate system (i.e., earth axis) the orientation of a second coordinate system (i.e., body axis). Any orientation of the second system is obtainable from that of the first by rotation through each of the three angles in turn, the sequences of which is important.

Europa.   The 4th largest satellite of Jupiter orbiting at a mean distance of 671,000 km. Its diameter is about 3100 km. Also called Jupiter II.

evection.   The departure of the Moon's orbit from an ellipse due primarily to perturbation by the Sun. The eccentricity is decreased during a little more than one half of a synodic period and increased for an equal time.

event horizon.   The effective surface of a black hole. The boundary around the black hole from within which matter and energy do not escape.

evolution, stellar.   The process of change from one stellar form to another. In general, the stars evolve from prestellar or protostellar contracting clouds to main sequence (q.v.) stars which derive energy by "burning" hydrogen (i.e., fusing it into helium by means of nuclear reactions). When main sequence stars run out of hydrogen in their central regions, they evolve to the red giant state, with contracted hot central cores "burning" helium and gigantic, expanded, cool, outer atmosphere. After the helium is exhausted, they evolve into more complex, short-lived states where heavier elements may be "burned" and still heavier elements created. Many of these states may be unstable and thus produce variable stars. The less massive stars may evolve relatively smoothly to white dwarfs, but more massive stars may blow off mass, and the most massive stars may produce supernova explosions— eventually producing neutron stars or black holes. More massive stars always evolve faster than less massive ones, since their central regions become hotter and "burn" nuclear fuels faster. A hot, massive, O-type star (about 20 solar masses or more) may last only a few million years on the main sequence, whereas a solar-type star (1 solar mass) may last 10 bllion years on the main sequence.

evolutionary theory of the universe.   Postulates that the present universe originated as a very highly condensed mass that at a particular time became unstable and suddenly started to expand. In the course of time (estimated to be a few hundred million years) the expanded materials aggregated into separate portions which were the protogalaxies (i.e., the precursors of galaxies). Contraction within the protogalaxies then led to the formation of galaxies and their constituent stars. The outward motion of the material initiated by the great expansion (the so-called "big bang") is still continuing and is apparent as the expanding universe. The galaxies that had the largest initial velocities are now the most distant from the local galaxy and from each other.

exit pupil (telescope).   The linear diameter of the virtual image of the telescope objective formed at the eyepiece (visible as a circle of light projected on the eyepiece lens). To take advantage of all light gathered by the objective during visual observation, the exit pupil should be smaller than the diameter of the eye's pupil, so that all the light passes into the eye.

exosphere.   (1) The outer Earth's atmosphere from about 1600 to 3000 km altitude where many atoms and ions at high temperatures move at speeds comparable to escape velocity and allow the occasional escape of gas into space. The region is composed essentially of protons, but is distinct from the interplanetary hydrogen; its temperature is some 1500° K. The exosphere is sometimes called the hydrogen geocorona. (2) In any planetary atmosphere, a high-altitude region where lighter, fast-moving atoms have a significant chance of moving faster than escape speed and escaping into space.

expanding universe.   A term popularized (but perhaps poorly chosen) by Eddington to describe the observation that the galaxies are receding from each other with the most remote galaxies receding fastest, as if galaxies were a swarm of sparks expanding from an explosion.

explement.   An angle equal to 360° minus a given angle. Thus 150° is the explement of 210° and the two are said to be explementary.

exterior planet.   Any planet whose orbit lies outside Earth's orbit.

exterior planet configurations and phases.   (See Fig., p. 85.)

extinction.   (1) The attentuation or dimming of light due to intervening haze along the light beam; that is, the reduction in illuminance of a collimated beam of light as the light passes through a medium such as interstellar dust wherein absorption and scattering occur. The dimming of starlight by the Earth's atmosphere or interstellar clouds; also the attentuation of radio radiation by the interstellar medium and the disk of the galaxy. (2) The amount of such extinction at a specified wavelength expressed in magnitudes.

extinction coefficient.   The amount of extinction per unit distance. In the galactic plane it averages about 1.9 magnitudes/kiloparsec and drops as low as 0.3 magnitudes/kiloparsec between dust clouds.

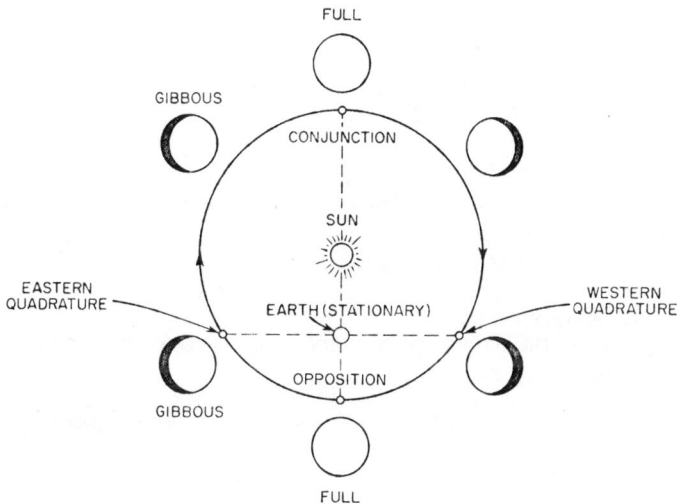

extragalactic.   Outside our galaxy, which is the Milky Way.

extragalactic nebula.   An obsolete term for a distant galaxy, derived before galaxies were proven to be star systems like the Milky Way, and not gaseous nebulae.

extrapolation.   Prediction of a quantity for which no observation is available (i.e., population in the year 2000) based on the trend of change of that quantity (i.e., population growth at mid-century) among data points that have been observed.

extraterrestrial radiation.   In general, solar and cosmic radiation received just outside the Earth's atmosphere.

extrinsic variable.   A variable star whose light changes are due to external factors and not to some property of the star itself (i.e., an eclipsing variable).

eye lens.   In an eyepiece, the lens closest to the eye.

eyepiece.   A small magnifier, usually of two or more lenses, placed behind the focus of a telescope (or microscope) to allow the eye to examine the magnified image. Various focal lengths of eyepieces give various magnifications.

eyepiece projection (telescope).   The system in celestial photography in which an eyepiece is used to enlarge the image before it falls on the film.

eye relief (telescope).   The distance the eye must be placed from the eye lens to obtain sharpest vision. In telescopes, binoculars, etc., large eye relief is important for users who wear glasses.

F

F corona. The outer solar corona; its light is sunlight scattered by interstellar dust.

F layer. A region (usually broken into two subregions) in the high ionosphere with high densities of electrons causing reflection of radio waves and permitting long distance communication over the line-of-sight horizon.

f number (or f ratio). The ratio of focal length divided by aperture for a telescope or lens. A measure of the system's photographic "speed" (the lower the number, the brighter the image).

faculae. Latin for "torches." Parts of the Sun's surface brighter than the general level of intensity of the photosphere and usually found near sunspots. Best visible near the solar limb, they are clouds of glowing vapor that float at an altitude of several hundred miles above the photosphere. In some mysterious way faculae seem to herald the outbreak of a new spot group; they also linger after the group has decayed. The faculae are really of about the same luminosity as the photosphere but since they are floating at a considerable altitude they are less effected by dimming, and so stand out clearly.

fall. A meteorite that was seen to fall (as opposed to having been found after an unseen fall).

fast. Adjective applied to an optical system with a focal ratio less than about 2 (cameras) to 4 (telescopes).

fast neutron. A neutron of 100,000 electron volts or greater.

field. (1) A region of space within which each point has a definite value of a given physical or mathematical quantity (i.e., gravitational field, magnetic field). (2) The area visible through a telescope.

field lens. The lens furthest from the eye in an eyepiece.

field stars. (1) Stars randomly scattered in space and not belonging to individual clusters. (2) Random stars appearing in a photograph of a selected object.

field of view. The angular diameter of the field covered by a telescope or other optical device.

figure of mirror. The shape of the surface of the mirror in a reflecting telescope. The mirror is ground or "figured" to within a few millionths of an inch of its desired theoretical shape. A sudden change in temperature during the night may distort the shape of the mirror and give a poor image and the figure of the mirror is said to be temporarily bad, since it assumes its visual figure upon reaching temperature equilibrium.

figure of revolution.   Three-dimensional figure produced by the revolution of a curve around its axis.

filaments, solar.   Dark linear markings shown on the Sun's disk by a spectrohelio-graph and caused by prominences seen in projection on the disk.

filar micrometer.   A micrometric measuring device used to measure small angular separations or diameters (i.e., double star separation, the distance of a comet from a known star, the angular size of a planet or a lunar crater, the angular distance of satellites from their primaries). The instrument consists of two fine filaments—one fixed and one movable—placed in front of the field lens of the telescopic eyepiece. Change of position of the movable filament is controlled by a calibrated drum and screw arrangement so the separation of the two filaments can be read to fractions of seconds of arc and thus indicate the separation of the components of the binary. If a series of position angles and corresponding distances is taken over a period of time, the orbit of the secondary around its primary becomes known.

find.   A meteorite found without having been seen to fall.

fireballs.   Exceptionally bright meteors. These objects are often called bolides, from the Greek word meaning "to throw." A fireball differs from a meteor in classification simply by the fact that it is brighter than mag. −3.5 (approximately). Fireballs are produced by pieces of cosmic material larger than those which give rise to the less brilliant meteors. If the chunk of material is large enough it produces a smoky trail, an illumination of the entire night sky, or explosive flashes.

First Quarter.   The phase of the Moon when it has completed one quarter of one complete revolution round the Earth relative to the earth–sun line. It appears as a half moon in the evening sky.

First Point of Aries.   The intersection point of the ecliptic and celestial equator reached by the Sun at the time of the vernal equinox (about March 21). It was designated several thousand years ago when it lay in Aries, but owing to precession of the equinoxes it is now in the constellation of Pisces. By definition this point has zero right ascension and zero declination.

first surface mirror.   A mirror in which the reflector coating is placed in the front rather than back surface.

fixed stars.   So distant are the stars that naked-eye observations of them over many years fail to reveal any changes in their relative positions. They therefore appear to be fixed in the heavens. The term is used popularly to designate the seeming immutability of the heavens though in reality the stars are not fixed, but ceaselessly moving.

Flamsteed number.   A number sometimes used with the possessive form of the Latin name of the constellation to identify a star (i.e., 72 Ophiuchi). The Flamsteed number is used for stars numbered in *Flamsteed's British Catalogue* of 1725. For stars which do not appear in Flamsteed's catalogue, numbers from other catalogues are used.

flares.   A bright, energetic region associated with a magnetic disturbance (usually a sunspot) on the Sun's surface. Once in a while a localized patch in a center of activity suddenly brightens in a few minutes and then dies away after an hour or so. Larger flares last longer, releasing as much as 10^{32} ergs of energy and expelling as much as 10^{16} grams of gas at velocities around 1500 km/sec. These chromospheric flares generally occur near the center of a group of sunspots; they are always associated with faculae, and 99 percent of them occur near a sunspot. Radiation and particles from large flares disrupt the Earth's ionosphere and can disrupt radio communications.

flare stars.   Main-sequence stars with flares lasting for hours and occurring erratically with no detectable periodicity. Best known are red dwarf (or UV Ceti) flare stars, which flare up from time to time in an irregular manner, somewhat like the Sun. Simultaneously increased radio emission from the star has been detected. The nearest flare star is about 8.6 light years away (i.e., about half a million times as far as the Sun from Earth).

flash spectrum.   An emission spectrum from the Sun's atmosphere, visible only during moments before or after a total eclipse. The characteristic feature of the chromosphere is the reddish color of the crescent of the solar surface that is visible just prior to totality in an eclipse of the Sun (hence the name chromosphere). Spectroscopic examination shows that at this time the dark-line Fraunhofer spectrum suddenly changes to a bright-line emission spectrum; this is called a flash spectrum because it lasts for only a few seconds and flashes out suddenly when the solar disc is hidden by the Moon. It arises in the hot, thin gases just above the Sun's surface.

flattening of the earth.   Distortion of the Earth by its spin causing the polar diameter to be about 42 km less than the equatorial diameter.

flocculi (also called plages).   Bright rims on sunspots revealed by spectroheliograph photographs taken in monochromatic light and originating in the chromosphere.

fluorescence.   The absorption of radiation of one wavelength, exciting electrons in atoms and causing re-emission of radiation at longer wavelengths as the electrons work their way back down to the initial energy state. A familiar example is the conversion of ultraviolet light to visible light in certain minerals. Fluorescence occurs in many nebulae where exciting (illuminating) star is hotter than 20,000° K. The nebula then absorbs the star's ultraviolent light and re-emits this energy in the form of emission lines. A good example is the Orion nebula, which has a bright-line spectrum and is associated with hot stars of class O and B.

flux.   (1) The amount of energy, usually in the form of radiation, light, or radio waves, passing through a unit area (such as 1 square centimeter or 1 square meter) per second. (2) A similar measure of flow of quantities other than energy, such as solar protons or meteorites.

flux unit. In radio astronomy, 10^{-26} watts per square meter per hertz, equaling 1 jansky.

focal length (telescope). For a single mirror or thin lens, the distance measured along its optical axis from the center of its surface to the focus, where the image of a very distant object (such as a star) is formed. In a compound telescope it is the distance from the entrance pupil to the focus and is called the effective focal length.

focal plane. The surface in which the image of very distant objects is formed, and lying perpendicular to the optical axis.

focal plane shutter. A shutter which moves in the focal plane (instead of at the lens) in order to expose the film in a camera system.

focal ratio. The "speed" of a lens or telescope mirror, expressed as the ratio of focal length to aperture diameter (lens size). A telescope of focal ratio f/4 has an aperture equal to one quarter of its focal length and can photograph a nebula or other extended object in one ninth of the exposure time required with an f/12 telescope.

foci. Two fixed points on the semiaxis major of an elliptical orbit, and equidistant from the center. In any system where a small body orbits around a substantially larger body (e.g., Earth-Moon system), the larger body lies at one focus and the other is unoccupied. In planetary orbits in the solar system, the Sun lies at one focus.

focus. The point to which light and heat are reflected by a concave mirror or refracted by a convex lens. The point where rays from a very distant object come together in an optical system.

Fomalhaut (γ Piscis Austrini). A southern 1st magnitude star (declination $-30°$). A main-sequence star of spectral class A, located only about 23 light-years away. Its visual magnitude is 1.2. Fomalhaut was known as "The Solitary One" because of its isolation from other bright stars. A 9th magnitude companion about 30″ away was found in 1896.

forbidden lines. Spectral lines arising from metastable (i.e., long-lived) energy levels in atoms (q.v.) and observed only in very thin (i.e., interstellar) gas. Because the lines can result only from spontaneous transitions out of these levels or orbits, they are not seen in dense gas (such as air) where atoms are disturbed by collisions with other atoms before the required transition can occur. Forbidden lines are designated by enclosing the source atom in brackets (i.e., [OH], a line from ionized oxygen).

fork mounting. A telescope mounting in which the tube is mounted between two prongs of a fork, the shaft of which is supported on bearings.

Fornax (the Furnace). An obscure group joining Cetus, added to the sky by the French astronomer Lacaille in 1752 and originally known as Fornax Chemica (the Chemical Furnace). The area is 398 sq deg. The main interest of this constellation centers in its conspicuous and relatively nearby galaxies.

Fornax galaxy. A small, spheroidal, elliptical galaxy about 3 kpc across and 170 kpc away with about 1 percent the mass of the Small Magellanic Cloud.

Foucault pendulum. A massive pendulum bob on a long cord used to demonstrate the rotation of the Earth. Below is the principle of Foucault's pendulum. At the poles, the Earth rotates beneath the plane of the pendulum oscillation. At the equator, the force of gravity is perpendicular to the axis of the Earth's rotation so the pendulum does not appear to rotate.

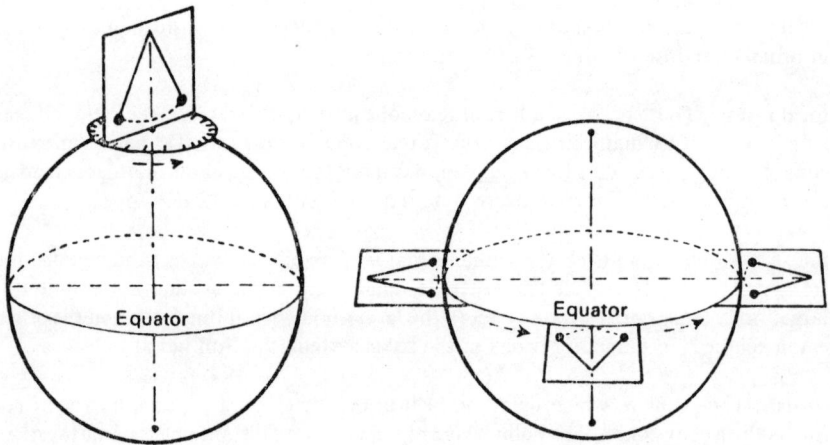

Foucault test. A method of testing telescope mirror surfaces by cutting reflected rays with a knife edge and observing the apparent shadings produced on the mirror. The eye, knife edge, and light are all located at the center of curvature of the mirror.

Fourier analysis. Analysis of a varying function by a mathematical technique representing the seemingly complex variations by the sum of a number of simple, periodic variations.

fractional method. A means of estimating the magnitude of a variable star by judging its brightness as a fraction of the interval between two comparison stars, one brighter and one fainter than the variable.

Fraunhofer lines. The more prominent absorption lines in the solar spectrum. The underlying solar spectrum is basically continuous, but is interspersed by thousands of sharp, dark lines of different intensities. These lines are called Fraunhofer lines after the Bavarian optician Josef von Fraunhofer who in 1814 detected over 500 dark lines in the Sun's spectrum. He designated prominent ones A, B, C, etc., in order of decreasing wavelength, but only a few designations (D, H, K) are still in use.

free atmosphere. That portion of the Earth's atmosphere above the planetary boundary layer in which the effect of the Earth's surface friction on the air motion is negligible and the air is usually treated (dynamically) as an ideal fluid. The base of the free atmosphere is usually taken as the geostrophic wind level.

free electron. An electron not bound to a specific atom. The neutral atom has its full quota of electrons, normally at the lowest possible levels. After collisions with photons or other particles, electrons may be knocked out of the atom and become free; the atom is said to be ionized.

free–free transition. A change in energy due to an encounter (without capture) between an atom (or ion) and an electron; bremmsstrahlung arises from such transitions.

free space. An ideal, perfectly homogeneous medium possessing a dielectric constant of unity in which there is nothing to reflect, refract, or absorb energy. A perfect vacuum possesses these qualities.

frequency. Number of periodic changes (cycles) per second. The period is the reciprocal of the frequency. Radio waves have frequencies of thousands of cycles per second (kilocycles per second or kc/sec) to many millions (megacycles per second or Mc/sec). Visible light's frequency is of the order $6(10^{14})$ cycles per second. For all electromagnetic radiation, frequency is given by c/W, where W is wavelength (cm) and c is the velocity of light, 3×10^{10} cm/sec. Hertz is a unit of frequency equal to one cycle per second.

fringes. The alternate dark and light lines caused by interference of electromagnetic radiation, seen in an interferometer (q.v.).

front surface mirror. See first surface mirror.

full moon. The Moon at opposition with a phase angle of $0°$. It appears as a round disk to an observer on the Earth because the entire illuminated side is toward Earth.

fundamental stars. Reference stars with very accurately measured positions.

G

gabbro. A coarse-grained igneous rock similar in composition to basalt (q.v.) which is fine grained. Unlike basalt gabbro is not erupted, but formed by the same magma type cooled underground. Gabbro is common on Earth and in the lunar uplands.

galactic cluster. See open cluster.

galactic coordinates. Celestial coordinates based on the appearance of the galaxy seen from the solar system. The galactic system of coordinates has as its physical basis the plane of symmetry of our galaxy. Its fundamental great circle is the galactic equator (a great circle fitted along the center of the Milky Way). The poles are the north galactic pole in Coma Berenices and the south galactic pole in Sculptor. Coordinates are called galactic latitude and galactic longitude. The point of origin (0° galactic latitude and 0° galactic longitude) is the direction to the center of our galaxy, which is located in Sagittarius. The plane of the galaxy makes an angle of 62.6° with the plane of the Earth's equator, and thus the angle between the two poles in the sky is also 62.6°.

galactic latitude. The distance in degrees from the galactic equator to any point on the celestial sphere measured on a perpendicular great circle.

galactic longitude. Galactic longitude is measured in degrees from the galactic center eastward to the point in question.

galactic plane. The plane defined by the galactic disk.

galactic poles. The two opposite points that are farthest north and south of the Milky Way. ($12^h49^m+27.^04$; $0^h49^m-27.^04$).

galactic wind. Outflow of hydrogen from the galactic center.

galaxies, barred spirals. Spiral galaxies with characteristic bar-like arms in the central regions. More than two-thirds of the recognized spiral galaxies are of the normal type. The remainder are barred spirals. Instead of emerging directly from the central region, the arms of this type begin abruptly at the extremities of a broad light bar that extends through the center. Barred spirals are arranged in Hubble's classification in a sense paralleling that of the normal type. The classes are designated SBa, SBb, and SBc. As the series of barred spirals progresses, the arms build up and unwind. In class SBa the arms are joined to form an elliptical ring, so that the galaxy resembles the Greek letter theta. In class SBb the ring is broken and the free ends are spread so that the galaxy is more nearly like the normal spiral. In class SBc the ends are so far separated that the galaxy has the form of the letter S.

galaxies (clusters).   Many and perhaps all galaxies occur in clusters that fill all known space. A few of the most prominent clusters are commonly known by the names of the constellations in which they appear; an example is the Virgo cluster.

Galaxy clusters, Their Approximate Distances and Velocities

Clusters	Distance (l.y.)	Velocity (1cm/sec)
Virgo	62,000,000	1,180
Pegasus I	210,000,000	3,700
Perseus	320,000,000	5,400
Coma	370,000,000	6,700
Ursa Major I	430,000,000	15,400
Leo	1,000,000,000	19,500
Gemini I	1,100,000,000	23,300
Bootes	2,100,000,000	39,400
Ursa Major II	2,200,000,000	41,000
Hydra	3,300,000,000	60,600

galaxies, elliptical.   The most abundant type of galaxy per unit of volume of space. Galaxies are so named because the nearer ones appear through the telescope as elliptical disks, lacking spiral arms. They are designated by the letter E followed by a number, which is 10 times the ellipticity of the disk. The series runs from the circular class E0 to the most flattened E7, which looks like a convex lens viewed edgewise. They range from the most massive to the least massive galaxies. They have little gas, dust, or young stars.

galaxies, irregular.   Galaxies lacking regular shape. They are designated by the letters Irr I for those with O and B stars and emission nebulae, and Irr II for those which cannot be resolved into stars. The best known irregular galaxies are the two nearby Magellanic Clouds, although the large cloud has also been called a barred spiral with one arm.

galaxies, local group.   The galactic system is a member of a group of at least 17 galaxies which occupy an ellipsoidal volume of space more than 2 million light-years in its longest dimension. Our galaxy and M31 are near the two ends of this diameter; these and M33 are the three normal spirals of the group. The Magellanic Clouds and 2 smaller systems have usually been classed as irregular galaxies. The remaining 10 are elliptical galaxies.

Designation	Type	Distance (Light Yrs)	Apparent Diameter	Linear Diameter (Light Yrs)
Milky Way	Sb	0	360°	98,000
Large Mag. Cloud	Irr	170,000	8°	23,000
Small Mag. Cloud	Irr	200,000	3°	10,000
Ursa Minor System	E	220,000	40′	3,000

Designation	Type	Distance (Light Yrs)	Apparent Diameter	Linear Diameter (Light Yrs)
Draco system	E	220,000	15′	1,000
Sculptor system	E	280,000	30′	2,000
Fornax System	E	550,000	40′	6,000
Leo I System	E	750,000	10′	2,000
Leo II System	E	750,000	8′	2,000
NGC 6822	Irr	1,500,000	15′	7,000
NGC 185	E	2,200,000	6′	4,000
NGC 147	E	2,200,000	9′	6,000
IC 1613	Irr	2,400,000	12′	8,000
M31	Sb	2,200,000	3°	120,000
M32	E2	2,200,000	5′	4,000
NGC 205	E5	2,100,000	12′	7,000
M33	Sc	2,100,000	35′	21,000

galaxies, spiral.   The most familiar type of galaxy, with spiral arms of stars and nebulae radiating from a bright center. The Milky Way is a typical spiral. The spiral galaxies are usually quite large, ranging from 20,000 to more than 125,000 light-years in diameter. The Milky Way and Andromeda are examples of the larger spiral galaxies. Spiral arms contain abundant dust, gas, and newly formed, bright, massive, hot, bluish stars often in clusters. Central regions have little gas and dust and are dominated by old, red giant stars. Spiral galaxies have the outlines of flattish lens-shaped disks with a maximum thickness at the center of roughly 10 to 15 percent of the diameter. From the calculated masses and brightness of the galaxies it is estimated that each contains from approximately one billion to one hundred billion or more individual stars.

galaxy.   A massive stellar system containing about 10^6 to 10^{12} solar masses of stars, gas, and dust. Different forms are believed to originate from different initial conditions of mass, rotation rate, turbulence, and other properties. The galaxies are believed to have formed roughly 13 billion years ago, some time after the big bang, as expanding primeval hydrogen broke up into separate clouds of galactic mass, each forming a protogalaxy. See galaxies, local group.

galaxy types.   The best known system of classification of galaxies was introduced by E. P. Hubble in 1925. It recognized four main classes: elliptical (E), ordinary spirals (S), and barred spirals (SB) and irregulars (I). Among spirals three subtypes, Sa, Sb, Sc, are distinguished according to the relative size of the nuclear or central bulge and the relative strength of the arms. Spiral galaxies show their typical spiral arms emerging directly from a bright round nucleus (ordinary spirals) or at the ends of a diametrical bar (barred spirals). Elliptical galaxies have a smooth structure from a bright center out to indefinite edges; they differ only in ellipticity from round (E0) to a 3:1 axis ratio (E7). Irregular galaxies are either of the Magellanic Cloud type, or chaotic and impossible to classify. The classification system was later refined to include new types or subtypes identified since 1925. Other classifications have been proposed. Morgan's system includes shape, orientation, and spectral type of the average (integrated) star light.

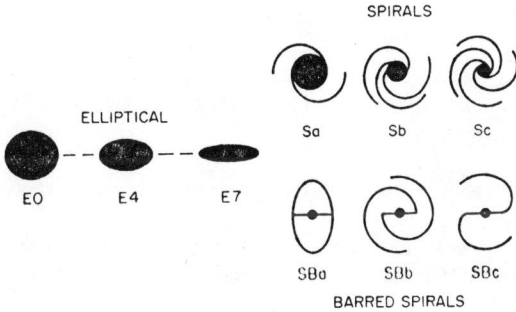

SPIRALS

ELLIPTICAL

Sa   Sb   Sc

EO   E4   E7

SBa   SBb   SBc

BARRED SPIRALS

Galilean refractor (telescope).   A "spyglass" type of refractor giving an erect image. It seldom has a magnifying power greater than 5 and usually has an aperture of less than 2 inches. Its deficiencies are its low resolving power, small field, and lack of image brightness.

Galilean satellites.   The 4 largest satellites of Jupiter, first reported by Galileo in 1610.

Galilean telescope.   A refracting telescope that uses a concave lens for an eyepiece. It produces an erect image.

Gamma Andromedae.   Known as Almach, it is considered one of the most beautiful double stars, with magnitudes 2.1 and 5.1 and contrasting orange and blue colors. They are separated by 10″. The brighter star is about 240 light-years from us.

Gamma Cassiopeiae.   Gamma lies in the girdle of Cassiopeia. It is a binary. The two companions are 2″ apart with magnitudes of 2 and 11. It is estimated to be 620 light-years away. A most peculiar star, subject to occasional "bursts" which send it up from its usual brilliance to well below the second magnitude (about 1 1/2).

Gamma Cygni.   Was called Sadir by the Arabs, the name meaning "the Hen's Breast." It lies at the center of the northern cross, with magnitude 2.2, at an estimated distance of 800 light-years. A low-power glass shows it to be located in a wonderful field of stars and immersed in a fine nebulous matter which extends over a wide area. The space between this star and Alberio is packed with the star masses of the Milky Way. They are so dense that Sir William Herschel counted over 300,000 within a distance of 5°.

Gamma Draconis.   The star Eltanin, whose observation in 1725 by English astronomer James Bradley led to the discovery of the aberration of light. It is about 120 parsecs away.

Gamma Leonis.   Gamma is the brightest star in the blade of the Sickle. It is called Algieba, meaning "the Forehead," from its location in the head of the Lion. It has been called the finest double in the northern sky; the components are bright orange and greenish yellow in color with magnitudes 2.6 and 3.8. They revolve around each other in a period of a little more than 400 years. The system is estimated to be 110 light-years distant.

Gamma Orionis.   Gamma, in the left shoulder of Orion, is known as Bellatrix, the name meaning "Amazon Star." It is of magnitude 1.6, pale yellow in color, and about 300 light-years away.

gamma ray.   Very short-wave, high-energy electromagnetic radiation often emitted by a nucleus. Each such photon is emitted as the result of a quantum transition between two energy levels of the nucleus. Wavelengths are generally less than a few tenths of an angstrom. Gamma rays have energies usually between 10 thousand electron volts and 10 million electron volts.

gamma-ray astronomy.   Astronomical observing based on detection of celestial gamma-ray sources. High-energy gamma rays are produced when a proton of high energy (such as is found in cosmic rays) strikes another proton (or any other nucleus). Short bursts of cosmic rays were first detected in 1967 by satellites monitoring the nuclear test ban treaty. Cosmic ray sources may involve high energy collisions between gas and dense stars or black holes.

Ganymede.   The largest satellite of Jupiter orbiting at a mean distance of 1,070,000 km. Its diameter is about 5000 km. Also called Jupiter III.

gaseous nebula.   Outmoded term for nebula, originating when some nebulae were thought to be stellar.

gas stream.   A stream of gas present in binary star systems in which one star has evolved and is ejecting mass which flows toward the other through the Inner Lagrangian point of the system.

gauss.   A unit of magnetic field strength; the force on a unit magnetic pole is 1 dyne where the field is 1 gauss.

Gaussian distribution.   A statistical distribution typical of errors in measurement and other variables clustered in a bell-shaped curve around a mean value.

Gaviola test.   A simplified test for Cassegrainian secondary mirrors.

GC (followed by a number).   The number of a star in the *General Catalogue* by Lewis Boss, which lists accurate positions and proper motions of 33,342 stars.

gegenschein.   The "counterglow," a part of the zodiacal light. The zodiacal light fades at large angular distance from the Sun, but brightens to a faint patch of light—

the gegenschein—at a point opposite the Sun. It is believed to result from zero-phase reflection of sunlight due to interplanetary dust.

gemination.   A term once used to describe the twinning of supposed Martian canals or the unpredictable transformation of a canal from a single line to two parallel lines.

Gemini (The Twins).   The 3rd constellation of the zodiac to which Ptolemy assigned 18 stars; in ancient times these twins were known by various names but are now known universally as Castor and Pollux, the sons of Leda, represented by the two bright stars. It is a conspicuous group stretching from these stars into the Milky Way. The modern area is 514 sq. deg. There are some fine double stars in Gemini including the brilliant and interesting system in Castor.

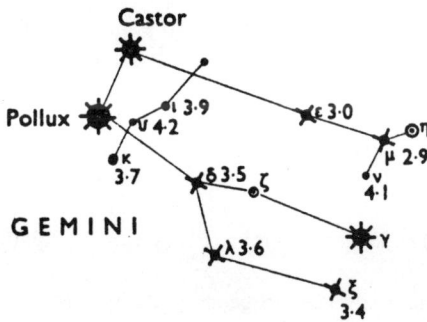

Geminids.   A rich meteor shower with maximum activity about December 12.

general precession.   The total precession caused by lunar, solar, and planetary gravitational forces on the Earth. It is a shift in the Earth's axis orientation with respect to the stars. This shifts the equinoxes westward along the ecliptic at the rate of about 50.3 seconds of arc per year. The effect of the Sun and Moon (lunisolar precession) produces a westward motion of the equinoxes along the ecliptic. The effect of other planets (planetary precession) produces a much smaller component motion in the other direction.

geocentric.   As seen from or referred to the center of the Earth. Relating to the Earth as a center or to the Earth's center of mass.

geocentric diameter.   The angular diameter of a celestial body measured in seconds of arc as viewed from the Earth's center.

geocentric latitude.   Of a position on the Earth's surface; the angle between a line to the center of the Earth and the plane of the equator.

geocentric parallax.   The angle between the direction of an astronomical object from a point on the surface of the Earth and the direction of the same object from the center of the Earth.

geocentric system.   As early as the 6th century B.C. the Earth was regarded by the Greek scholars as a stationary globe. The sky was thought to be a hollow concentric globe in which the stars were set, supported on an axis through the Earth; it rotated daily from east to west, causing the stars to rise and set. Within the sphere of the stars the Sun, Moon, and 5 brighter planets shared in the daily rotation. They also revolved eastward around the Earth, periodically retreating toward the west against the turning background of the constellations. The geocentric system remained almost unchallenged for more than 2,000 years, amplified meanwhile in attempts to account for the retrograde movement of the planets.

geocorona.   The ionized, gaseous envelope of the Earth—primarily hydrogen. This halo extends to about 15 Earth radii.

geodesic.   The curve on a surface joining two points whose length between these points is less than that of any curve on the surface joining them.

geodesy.   A branch of surveying concerned with the measurements of terrestrial gravitational forces, size and shape, land masses, and shapes of bodies in space by analysis of gravitational forces associated with those bodies.

geographical mile.   The length of 1 minute of arc of the equator, or 6,087.08 feet.

geoid.   A surface commonly used in geodesy, defined as the equipotential surface of gravity plus inertial forces, i.e., a surface such as a hypothetical universal sea level on which no work (or force) is required to move a mass from one point to another. The shape of the geoid is not smooth but has many undulations which are determined by the mass distribution on Earth below the surface. The essential characteristics of an equipotential surface, such as the geoid, is that the same amount of energy is required to move a given mass to infinity from any point on this surface. The term can be generalized to other planets. See also equilibrium spheroid (an equivalent concept).

geomagnetic coordinates.   A system of spherical coordinates based on the best fit of a centered dipole to the actual magnetic field of the Earth.

geomagnetic equator.   The terrestrial great circle everywhere 90° from the geomagnetic poles.

geomagnetic field.   The magnetic field surrounding the Earth out to a distance defined by its interaction with the solar wind. (See Fig., p. 99.)

geomagnetic latitude.   Angular distance from the geomagnetic equator measured northward or southward through 90° and labeled N to S to indicate the direction of measurement.

geomagnetic meridian.   The meridianal line of a geomagnetic coordinate system.

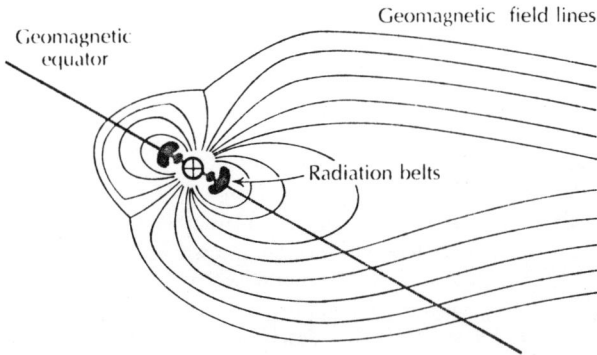

geomagnetic pole.   Either of two antipodal points making the intersection of the Earth's surface with the extended axis of the magnetic field of the Earth. The sites of vertical magnetic field lines.

geomagnetic storms.   Unusual agitations of the Earth's magnetic field indicated by erratic gyrations of the compass needle. They occur when streams of ions from the solar disturbance arrive on Earth a day or more after a flare is seen. These effects are sometimes accompanied by strong, induced earth currents which can interfere with wire communications. The incoming protons also cause the primary glow of the aurora in the upper atmosphere.

geomagnetism.   The magnetic phenomena exhibited by the Earth.

geometric albedo.   The fraction of light reflected from a planet relative to that reflected from a perfectly reflecting disk with Lambert surface properties at zero phase angle.

geosphere.   The solid and liquid portion of the Earth; the lithosphere plus the hydrosphere.

geostropic wind.   The wind blowing along straight isobars that produces a Coriolis force that will just balance the existing pressure gradient. It neglects centrifugal force.

German mounting.   An equatorial mounting in which two axes meet each other in a tee.

ghost image.   Nebulous areas of light that can be seen in certain types of eyepieces.

giant planets.   The planets Jupiter, Saturn, Uranus, and Neptune. Also called Jovian planets.

giant star.   A star of large size and luminosity. Giant stars represent stages post-main-sequence evolution after most hydrogen has been burned in the star's central core.

gibbous.   Phase of Moon (or any other planet or satellite) when more than half of the disk is illuminated, but the phase is less than full.

Giedi.   See Alpha Capricorni.

Giga.   A prefix indicating 10^9 (one billion).

globular clusters.   Groups of stars ranging in number from several thousand to a few million. Over 100 of these clusters are known in the Milky Way. Globular clusters have also been detected in external galaxies. The name arises from the almost symmetrical clustering of the stars in the form of a spheroidal mass. The diameters of globular clusters lie mostly between 70 and 600 light-years (20 and 190 parsecs), the average being approximately 200 light-years (60 parsecs). They are very old objects, with ages around 12 to 13 billion years. They are composed of Population II stars having very few heavy elements.

globule.   A very small dark cloud of dust, obscuring background stars. Typical dimensions are a fraction of a light-year. They are often called Bok globulars, after their discoverer.

Gnomonic Projection.   A projection in which the meridians and parallels of latitude are projected on to a plane tangent to the Earth (or other planet) at one point (i.e., the north pole). Meridians appear as straight lines toward the pole and parallels of latitude are not parallel; distortion increases with distance from the tangent point and great circles appear as straight lines.

Golay cell.   A tiny aneroid barometer whose internal pressure increases when heated by infrared radiation. The sensitivity of this device depends on its speed of response. Rapid interruptions of the incoming beam of radiation by means of a rotating sector produce pulsations that can be turned into an alternating current and amplified electronically. This detector has been used for very short radio waves, with wavelengths of the order of 0.1 mm.

Gould belt.   A system of stars and gas in our local part of the Milky Way (within about 300 pc) and tilted at about 10° to 20° to the galactic plane. It contains many of the most prominent O and B stars.

gradient.   A vector which measures the direction and magnitude of the greatest rate of decrease of a function. Examples include temperature gradients (°K/km) measuring temperature change with descent into the Earth or ascent into the atmosphere.

grains.   Micron-dimension (about 10^{-4} cm) solid particle in interstellar space or in clouds surrounding certain stars, such as suspected newly forming stars.

granules.   Small, bright features of the photosphere of the Sun covering 50 to 60 percent of the surface. They have a mottled appearance and have been likened to rice grains.

graticule. (1) The network of lines representing parallels and meridians on a map, chart, or plotting sheet. (2) A scale at the focal plane of an optical instrument to aid in the measurement of object.

grating. A set of parallel narrow slits or lines ruled on a glass plate which deflect light passing through them by the process of diffraction. A coarse grating used in front of a telescope lens gives several images of each bright star. A fine diffraction grating produces the spectrum in colors of light passing through it or reflected from a reflection grating. Gratings are the basic optical element in many modern spectroscopic devices.

gravipause. The boundary at which the gravitational attraction of two celestial bodies is equal.

gravisphere. The sphere within which the force of gravity of a given celestial body is the dominant gravitational force.

gravitation. The force which manifests itself as a mutual attraction between masses. It is a force of physical attraction which pervades the entire universe, binds all celestial bodies in a common entity, and makes all their actions interdependent. The force F between two bodies of mass M and m, separated by the distance r is $G\,Mm/r^2$, where G is the constant of gravitation. $G = 6.67 \times 10^{-8}$ in cgs units.

Gravitation, Law of. First enunciated by Sir Isaac Newton. It states: "all bodies tend to attract one another along the line connecting them with a force proportional to the product of their masses and inversely proportional to the square of the distance between them." See gravitation, where the mathematical formulation is given.

gravitational collapse. The relatively rapid contraction of a cloud or star when inward mutual gravitation of the constituent particles dominates over the outward pressure of the gas. Energy is radiated from the body during the collapse. See Helmholz contraction.

graviton. A hypothetical subatomic particle associated with gravity. It has no mass, no charge, and travels at the speed of light.

gravity. The result of the Earth's attraction (gravitation) directed nearly toward its center and diminished by the lifting effect of the Earth's rotation. The acceleration of gravity (g) is the rate at which a falling body picks up speed; its value at sea level increases from 978.039 cm/sec^2 at the equator to 983.217 cm/sec^2 at the poles, since the poles are closer to the Earth's center.

gray atmosphere. A model atmosphere constructed under the assumption that absorption of light occurs but is the same at all wavelengths.

gray body. A body with constant albedo (and emissivity) that is less than unity at all wavelengths.

great circle.   The circle on a sphere whose plane passes through the center of the sphere.

great distance.   The distance between any two points, measured in either degrees or miles along the great circle connecting them.

Great Red Spot.   See red spot.

great year.   The period of one complete cycle of the equinoxes around the ecliptic (about 25,800 years).

greatest elongation.   The maximum angular distance between an inferior planet and the Sun as observed from the Earth.

Greek alphabet.

	Capital	Lower Case		Capital	Lower Case
Alpha	A	α	Nu	N	ν
Beta	B	β	Xi	Ξ	ξ
Gamma	Γ	γ	Omicron	O	o
Delta	Δ	δ	Pi	Π	π
Epsilon	E	ϵ	Rho	P	ρ
Zeta	Z	ζ	Sigma	Σ	σ
Eta	H	η	Tau	T	τ
Theta	Θ	θ	Upsilon	Υ	υ
Iota	I	ι	Phi	Φ	ϕ
Kappa	K	κ	Chi	X	χ
Lambda	Λ	λ	Psi	Ψ	ψ
Mu	M	μ	Omega	Ω	ω

green flash.   A momentary green coloration of the last bit of the setting sun as it sinks below the horizon. The coloration is due to diffraction of the Sun's light by the atmosphere. It requires a distant, clear horizon. It may also be seen at sunrise. At the South Pole before the winter night, the flash is visible for half an hour.

Greenwich Civil Time.   The local twenty-four hour mean time at Greenwich, England; usually called Greenwich Universal Time. Used as a reference time system for observers throughout the world.

Greenwich meridian.   The meridian of longitude that passes through Greenwich, England. It is the origin for longitude by common usage and agreement.

Gregorian Calendar.   See calendar, Gregorian.

Gregorian telescope.   A compound telescope that makes use of a concave secondary mirror.

grit.   Term used by telescope makers to refer to abrasives.

ground state.   In an atom, the state with all the electrons in their lowest possible energy levels or orbits.

Grus (The Crane).   A southern constellation introduced by Bayer in 1604; it lies south of the bright star Fomalhaut and is conspicuous because of an arc of stars concave towards the second magnitude α Gru. The area is 366 sq deg. Many galaxies are found included in this constellation.

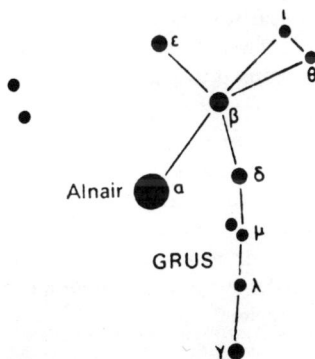

guided camera (telescope).   A camera attached to an equatorial mounting for the purpose of photographing celestial objects.

Gum nebula.   A network of faint filaments that span about 40° on the southern hemisphere. These filaments are related to the supernova remnant found in the constellation Vela, near the center of the area marked by the Gum nebula. The large angular extent indicates a supernova explosion some 11,000 years ago located fairly close to the solar system—maybe some 1500 light-years away. A pulsar with a period of some 10^{-2} sec located near the center is believed to be the supernova remnant. The nebula was discovered in the 1950's with wide-angle photography by Colin Gum.

H

H⁻ ion. An H atom with an extra electron—forming a negative ion consisting of a proton and two electrons. H⁻ is a major source of opacity in the Sun and other stars of the spectral type later than A5.

H I regions. Interstellar regions of neutral (unionized) hydrogen. The gas is not ionized because there are no nearby stars hot enough to produce the ultraviolet light required to ionize hydrogen. Twenty-one centimeter radio emission comes from netural hydrogen and has been used to map vast H I clouds.

H II regions. Emission nebulae or regions of ionized hydrogen gas. They are generally created in regions where material is heated by hot nearby stars whose ultraviolet light ionizes hydrogen. Temperatures range from 8000°K to 12,000°K, and up to 20,000°K. In line with a terminology that distinguishes between singly ionized (II) and neutral stars (I), areas where hydrogen is ionized are labeled H II regions. At these temperatures most solid particles evaporate, and atoms and ions are the most abundant components of an H II region.

H & K lines. Two lines of singly ionized calcium, at 3968 and 3934 Å, respectively. The spectrum of sunlight observed visually is an array of colors from violet to red, interrupted by thousands of dark absorption lines. The German optician Fraunhofer was the first, in 1814, to see the absorption lines clearly; he mapped several hundred dark lines and labeled them with letters beginning at the red end of the spectrum. Some lines including H and K are still known by the letters which he assigned to them. They are the strongest solar lines.

H-R diagram. A useful representation of stellar observational properties derived largely from the publications of the Danish-born Ejnar Hertzsprung in 1914. H-R is named for Hertzsprung and the American astronomer H. N. Russell. H. N. Russell did similar work independently, plotting spectral class vs absolute magnitude on a graph in which the horizontal axis was made up of the spectral classes in order of descending temperature, starting with spectral class 0 on the left and ending with spectral class M on the right. The absolute visual magnitude of the stars is plotted on the vertical axis, with more luminous stars toward the top. Theoreticians often use a form of the same diagram with surface temperature on the horizontal axis and Sun temperature on the vertical axis. (See Fig., p. 105.)

Hadar. See Beta Centauri.

hadron. Any subatomic particle associated with strong interactions inside atomic nuclei.

H-R diagram

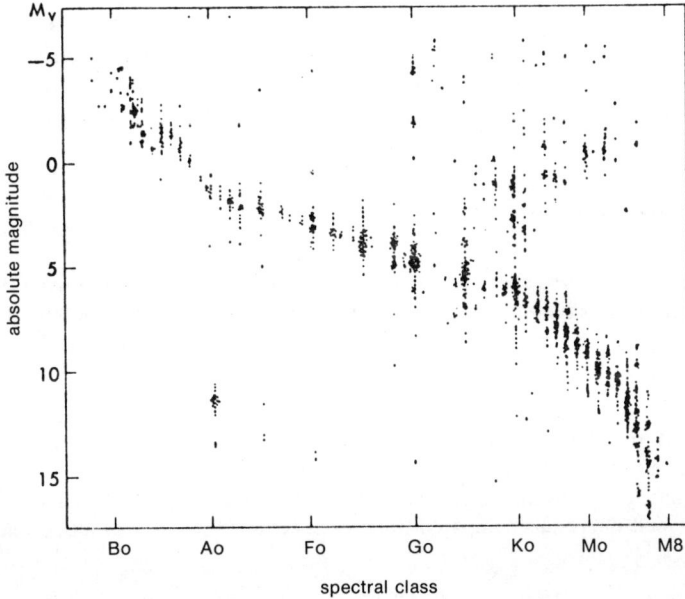

hail.   Ice balls or stones, with diameters ranging from a few mm to 5 cm or even more, either detached or fused in irregular lumps. They are either quite transparent or composed of alternating clear and opaque, snow-like layers, the clear layers being at least 1 mm thick. Hail occurs almost exclusively in violent or prolonged thunderstorms and never with temperatures below freezing at the ground.

halation.   The appearance of a star as a disk (or a disk surrounded by a ring) in a long-exposure photograph due to its light spreading into the photographic emulsion and reflecting off the back surface of the photographic plate.

half-life.   The time period within which half of any initial population will undergo some type of demise associated with the given population. An example is disintegration of radioactive isotopes, each of which has its own half-life.

Halley's Comet.   See Comet, Halley's.

halo.   (1) A glowing ring around the Sun or Moon caused by minute particles of ice suspended in the air. An outer, fainter ring of light may appear around the main halo, and "parahelia" or "sun dogs" (colored disks of light) may appear on either side of the Sun and 22° away from it. (2) The swarm of globular clusters and related objects in a roughly spherical volume around the Milky Way or other galaxies.

halo stars. Stars moving in orbits well above or below the galactic disk. When careful investigations were made of single stars in our galaxy with exceptionally high velocities (up to several hundred kilometers per second relative to the Sun and its neighbors), these objects were found to move outside the disc. Halo stars are typical metal-poor members of Population II (q.v.), which relates them to the stars in globular clusters lying outside the disk. Our galaxy is thus made up of two basic structures: the disc with its old and young stars and open clusters including a large reservoir of gas and dust, and the halo with its globular clusters and individual halo stars moving in isolation through the void.

Hamal. See Alpha Arae.

Harmonic Law. The cubes of the mean distances of any two planets from the Sun are to each other as the squares of their periods of revolution about the Sun; Kepler's third law, $a^3 = P^2$ where a is the distance from the Sun in A.U.s and P is the sidereal period in years.

Hartmann test. An optical test used primarily for large mirrors where the Foucault test is less effective. The mirror is tested by means of a screen in which holes are punched along various diameters. The holes are then photographed inside and outside focus and the results compared to find the actual focal point of any given zone.

Harvard sequence. A sequence of stellar spectra arranged in order of decreasing surface temperature based upon photographic spectra taken with an objective prism.

harvest moon. The full moon which falls on or nearest to the date of the autumnal equinox. The autumnal Moon when for several nights the full or nearly full Moon rises at approximately the same time early in the evening. Near the autumnal equinox in northern latitudes the delay in rising each day is about 13 minutes instead of the average 52 minutes.

Hayashi track. An evolutionary path followed across the H-R diagram by a star in the early convective stages of its pre-main-sequence evolution.

Hayford Spheroid. The solid figure with proportions the same as the Earth; a slightly flattened sphere with flattening amounting to less than 0.5 percent of the diameter.

haze, damp. Small water droplets in the atmosphere with the horizontal range of visibility usually considerably more than 1 1/2 miles. It is similar to a very thin fog, but the droplets or particles are more scattered than in light fog and presumably also smaller. This phenomenon is usually distinguished from dry haze by its grayish color, the "greasy" appearance of clouds seen through it (as though viewed through a dirty window pane), and the generally high relative humidity. Commonly observed on seacoasts and in southern states.

haze, dry.   Dust or salt particles which are dry and so extremely small that they cannot be felt or discovered individually by the unaided eye; however, they diminish the visibility and give a characteristic smoky (hazy and opalescent) appearance to the air. This phenomenon produces a uniform veil over the landscape and subdues its colors. This veil has a bluish tinge when viewed against a dark background such as a mountain, but has a dirty yellow or orange lint against a bright background such as the Sun, clouds at the horizon, or snow-capped mountain peaks.

HD (followed by a number).   The Henry Draper Catalogue number referring to stars with spectral type determined at the Harvard College Observatory. The H-R is a revision to the HD.

heat.   A form of energy whose quantity is measured in the change of temperature produced. It is, fundamentally, the kinetic energy associated with movements of molecules and atoms, which increases as temperature increases.

heat transfer.   Transfer of energy as heat (thermal energy) from one location to another. Heat transfer may occur in three different ways: radiation, conduction, or convection, or in a combination of these ways. Radiation is the transfer of thermal energy directly as electromagnetic radiation, which has a net flow from warmer to cooler bodies, heating the latter. Conduction is the transmission of heat by body contact (i.e., heating a pan or metal rod by placing it in a fire). Convection is the transfer of heat by currents in fluids or gases. Air is a gas and transfers much of its heat in this manner.

heavy cosmic-ray primaries.   The positively charged nuclei of elements heavier than hydrogen and helium up to the atomic nuclei of iron.

heavy elements.   A vaguely defined term referring to elements heavier than a certain limit, usually helium or boron.

heilegenschein.   A diffraction–reflection effect seen directly opposite the Sun around the shadow of the observer. It is often seen around the shadow of an airplane containing the observer, or in early morning or late evening on a moist, grassy lawn.

Hektor.   Trojan asteroid 624. It is the largest Trojan, estimated to be about 100×300 km. It may be a binary asteroid (separated or in contact) resulting from interaction of two Trojans.

heliacal rising or setting.   The first rising of a star after invisibility due to proximity to the Sun, or the last setting preceding its proximity to the Sun.

helio.   Prefix referring to the Sun.

heliocentric parallax.   The difference in the apparent positions of a celestial body outside the solar system as observed from the Earth and Sun. It can be derived by observing the body's shift in apparent position as Earth moves around the Sun.

heliographic.   Referring to positions on the Sun measured in latitude from the Sun's equator and in longitude from a reference meridian.

heliostate.   A clock-driven mirror (or coelostat) used to observe the Sun.

helium.   An element whose nucleus contains two protons. The most common form contains two neutrons, so that most helium atoms are about four times as massive as hydrogen. Helium is the second most abundant element in the universe and was discovered in the Sun.

helium flash.   The beginning of rapid, runaway helium-burning inside a star after the hydrogen in the central core has been exhausted. This exhaustion is followed by a contraction of the interior, increasing the pressure and temperature to a level of about 100 million degrees, whereupon helium begins to be consumed in highly energetic nuclear reactions that expand the outer layers of the star and turn the star into a giant.

Helmholtz contraction.   A form of gravitational collapse. It is a contraction of a star or cloud during which the internal temperature rises and energy is radiated. Once thought to be the energy source of the Sun, it is now thought to be a late stage of pre-main-sequence evolution of stars. It may be the main energy source in some pre-main-sequence or infrared stars. It is much slower than free-fall collapse, and terminates when nuclear reactions start in the star's interior.

Henry Draper classification.   Classification of stellar spectra into groups designated by the letters O, B, A, F, G, K, and M. These classes contain most of the stars. They do not fall exactly in alphabetical sequence because the classes were first assigned in order of hydrogen absorption-line strength and later rearranged in order of temperature. Also, some letters were assigned to classes that later had to be dropped. Much of the original work on the sequence was done by Miss Annie J. Cannon of Harvard, who showed that the system was capable of great refinement and achieved finer classifications by the use of numerical subclasses from 0 to 9. This solution allows in principle for a total of seventy classes ranging from O0 to M9.

Hercules (The Kneeler).   Hercules, the roman name of the Greek Hero, is the fifth constellation in order of size and occupies 1,225 sq deg between Ophiuchus on the south and Draco on the north. Containing a number of scattered stars from the 3rd magnitude downward, Hercules forms no particular pattern by which it may be recognized. The most interesting objects are the two globular clusters M13 and M92, and the red variable α (Rasalgethi). (See Fig., p. 109.)

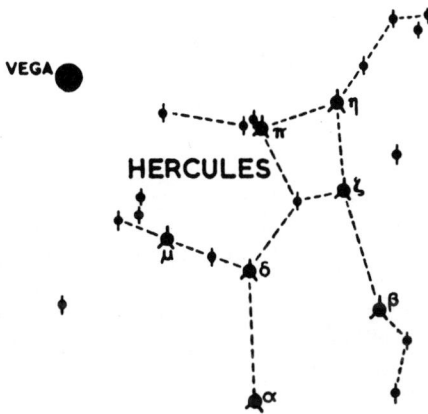

Herschel wedge.   A wedge-shaped prism used for observing the Sun. It permits only a small fraction of the sunlight to reach the eyepiece.

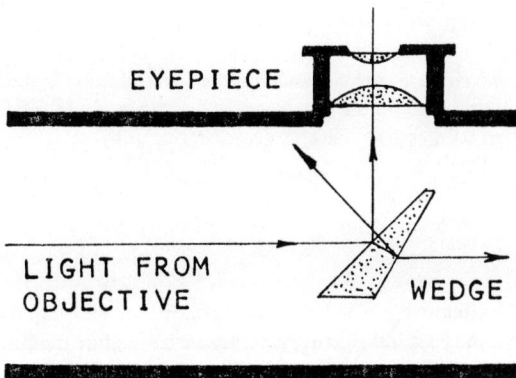

Hertz (Hz).   A unit of frequency equal to 1 cycle per second (cps). Radio frequencies are often given in megahertz (MHz) or millions of cycles per second.

Hertzsprung gap.   The region in an H-R diagram around zero absolute magnitude and spectral classes A0 to F5 where few stars seem to be present. The region is believed to represent an unstable stellar structure. The few stars known to lie in the region are variable.

H-H process.   See proton–proton chain.

Hidalgo.   An asteroid (944) with the very large orbital eccentricity (0.66) inclination (43°) and semimajor axis (5.8 A.U.).

high.   An atmospheric system characterized by relatively high pressure at its center and fair weather.

high velocity star.   A star with a high velocity (greater than about 120 km/sec) with respect to the Sun or local stars. They are mostly halo stars (q.v.).

Hind's Crimson Star (R Leporis).   A famous variable star ranging from magnitude 6 to 11 with a period of 436 days. In a telescope it is said to resemble a drop of blood. It has the deepest color of any readily observable star.

Hipparchus (Second Century b.c.).   Often called the "Father of Astronomy." Practically all our knowledge of him has been obtained secondhand from later astronomers. He measured the distance to the Moon in terms of size of Earth by an ingenious use of an eclipse, made the first recorded catalogue of the stars, and contributed to the theory of the motion of the planets. His greatest discovery was the precession of the equinoxes. Some of his work may have been based on still earlier astronomy, now lost.

Hirayama families.   Groups of asteroids with similar orbits suspected to be fragments from collisions between asteroid pairs.

horizon.   The great circle of the celestial sphere located midway between the zenith and nadir, or a line resembling or approximating such a circle. The place where the sky and Earth seem to meet (in the absence of mountains).

horizon system of coordinates.   A set of celestial coordinates, usually altitude and azimuth, based on the celestial horizon as the primary great circle and the zenith as a pole.

horizontal branch.   Stars in an H-R diagram (usually of a globular cluster) representing a range in spectral types but sharing roughly zero absolute magnitude. Giant stars lie along this branch.

horizontal parallax.   The geocentric parallax of a celestial body on the observer's horizon. This is equal to the angular radius of the Earth as seen from that body.

horizontal stratification.   Uniform meteorological conditions at a given altitude over the area under consideration.

Horologium (The Clock).   One of the southern constellations added by Lacaille in 1752. It is an inconspicuous group of area 249 sq degrees. Horologium lies well removed from the Milky Way and contains no galactic clusters or gaseous nebulae. There are no double stars of note, but there is an interesting globular cluster and a number of galaxies.

horse latitudes.   The region of comparatively light westerly winds found in the subtropical high-pressure belt.

Horse Head Nebula (NGC 2024).   A dark nebula in Orion, not far from Zeta Orionis in the Belt, silhouetted against bright background nebulosity. It is roughly 1100 light-years away.

hour angle.   Difference between sidereal time and right ascension, the time elapsed since the meridian passage of a celestial body. If the body is still in the eastern part of the sky, its hour angle is negative.

hour circle.   A great circle drawn from the pole through a star or some other point on the sky perpendicular to the celestial equator. It is a line of constant right ascension.

Hubble constant.   The constant relating the velocity of receding galaxies to their distance, designated H and having a value of 56.9± 3.4 km/sec/megaparsec. This means that for every megaparsec greater distance, galaxies are receding on the average of 56.9 km/sec greater speed. The increase of radial velocity v, with distance D, in the Hubble law of red shifts, or the constant H in the equation $v = HD$.

Hubble's nebula (NGC 2261).   The cone-shaped variable nebula extending from the infrared (young?) star R Monocerotis.

humidity.   In general, the moisture content of the atmosphere.

Hunter's moon.   The first full moon after the Harvest Moon. Like the Harvest Moon it rises with little delay from night to night, hence the name.

hurricane.   An extremely violent cyclonic storm of the tropics. It is characterized by a calm central eye a few kilometers in diameter in which the atmospheric pressure is very low, and a surrounding cyclonic vortex of great intensity. The hurricane is usually accompanied by torrential rains. It originates in regions near (but not on) the equator and moves rather slowly westward and poleward until it reaches about latitude 25° to 30° when it "recurves" toward the east as its speed increases.

Huygenian eyepiece.   An eyepiece consisting of two plano-convex lenses (or convex meniscus lenses) with the convex surfaces toward the objective. It has a large angular field, good eye relief, and is free from distortion. In its original form the Huygenian eyepiece was called a 4-3-2 lens (the numbers indicating the ratios between the focal length of the field lens, the distance between lenses, and the focal length of the eye lens in that order).

Hyades (Taurus).   An open cluster of about 200 stars. The bright stars in a V-pattern outline the face of the Bull with Aldebaran as its eye. These six naked-eye stars attracted much attention among ancient peoples and were associated with rainy and inclement weather. The cluster is about 500 million years old and about 130 light-years away. (See Fig., p. 112.)

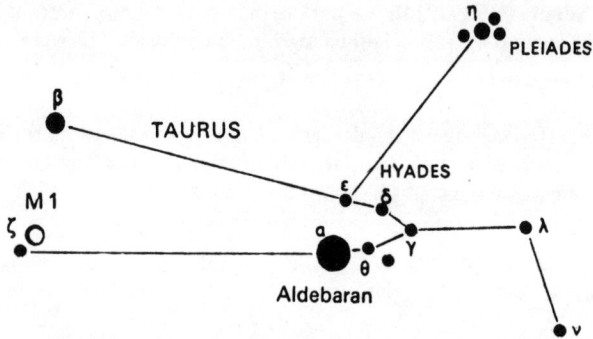

Hydra (The Watersnake).   The largest constellation in the sky, it is remarkably barren containing only one bright star—α, or Alphard, the Solitary One, clearly reddish with a magnitude 2.0. Castor and Pollux point toward Hydra. It was named after Hydra, the nine-headed serpent or monster of Lake Lerna slain by Hercules. It is an area of 1,303 sq deg. It is one of the old groups to which Ptolemy assigned 25 stars in all. It may have been designed around 2600 B.C. to mark the celestial equator of that era.

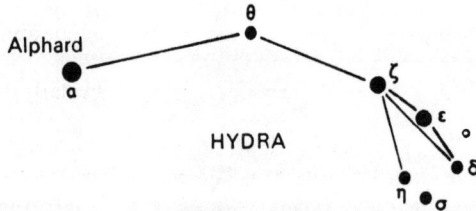

hydrogen alpha (Hα).   The most prominent spectral line of the Balmer series at wavelength 6563 A; H II region and other astronomical features are detected by their Hα emission.

hydrometer.   An instrument designed to measure density of liquids.

hydrosphere.   That part of the Earth consisting of liquid water: the oceans, seas, lakes, and rivers.

Hydrus (the Little Snake).   A constellation introduced by Bayer about 1603 between Horologium and Tucana. There are no objects in the constellation of special interest, but β Hydra is the nearest reasonably conspicuous star to the south celestial pole. The area is 243 sq deg. (See Fig., p. 113.)

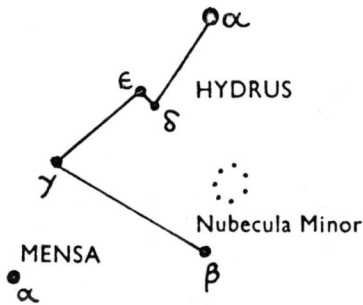

hygrometer. An instrument designed to measure and indicate the atmospheric humidity.

hyperbola. The curve followed by a body moving faster than escape velocity.

hyperboloid. A three-dimensional figure obtained by revolving a hyperbola around its axis.

Hyperion. A satellite of Saturn with orbital distance 1,483,000 km and diameter estimated to be about 500 km.

hyperon. An unstable, heavy subatomic particle with lifetime about 10^{-8} to 10^{-10} sec.

hypobaric. Pertaining to low atmospheric pressure, particularly the low atmospheric pressure of high altitude.

I

IAU. See International Astronomical Union.

IC (followed by a number). The catalogue number of a faint nebula, star cluster, or galaxy as listed in the *Index Catalog* (a supplement to the *New General Catalog*).

ISU. See International System of Units.

Iapetus. An outer satellite of Saturn with an estimated diameter of about 1200 km and an orbital distance of 3,560,000 km. With one face constantly toward Saturn, Iapetus is striking for its astonishing difference in albedo between the leading and trailing hemispheres. The leading side is only about 1/6 as bright as the trailing side. The leading side may be covered with dark, rocky soil and the trailing side with snow or ice. This may have occurred because of bombardment of the leading side by high-velocity meteorites, which may be particles knocked off the outer-retrograde satellite Phoebe. They could collide with the leading side of Iapetus as they spiral in toward Saturn under the influence of Poynting-Robertson forces (q.v.).

Icarus. Asteroid 1566, which is on an earth-crossing orbit and can approach within 6 million km. It has a diameter 1.1 km, a rotation period of 2^h16^m, and a relatively spherical shape. It has a semimajor axis of 1.07 A.U., but because of its high eccentricity it can come within 0.19 A.U. of the Sun—inside Mercury's orbit.

ideal gas (or perfect gas). A theoretical gas consisting of molecules represented by mathematical points with zero volume whose properties are extremely well simulated by most ordinary gases under a wide range of temperatures and pressures. Opposed to a degenerate gas in which molecules and atoms strongly interact and electrons are freed from atoms. It obeys the ideal gas law, q.v.

ideal gas law. The equation relating pressure P, density ρ, molecular mass m, and temperature T for an ideal gas (and common gases under ordinary conditions). $P = \rho k \, T/m$, where k = Boltzmann constant $- 1.38 \times 10^{-16}$ erg deg^{-1} kelvin^{-1}.

image tube (or image intensifier). The name given to a whole group of electronic imaging devices using a photocathode and producing an image with brightness enhanced over that of the original optical image. A television camera is such a tube. Light strikes a photosensitive surface, producing electrons multiplying the signal strength. An image tube having one or more electron multiplication stages is called an image intensifier; a tube having a cathode sensitive to certain wavelengths but producing an image at a different wavelength is called an image converter.

immersion. The disappearance of a body when occulted.

114

inclination. The angle between the plane of an orbit and some fundamental plane such as the plane of the sky (used with binary stars), or the ecliptic plane (used with solar-system objects).

inclination of the Earth. The tilt of the Earth's axis in relation to the plane of the Earth's orbit. The angle of inclination is 23 1/2° from the normal to the plane. More properly called obliquity.

index of refraction. A measure of the ability of a medium to refract lights, equal to the sine of the angle of incidence divided by the sine of the angle of refraction.

Air, dry, 0° C	1.00029
Alcohol, ethyl	1.36
Carbon disulfide	1.63
Carbon tetrachloride	1.46
Diamond	2.42
Glass, crown	1.52
Glass, flint	1.61
Quartz, fused	1.46
Water	1.33

Indus (The Indian). Another of the inconspicuous southern constellations defined by Bayer about 1603. The major star, α Ind, is an orange star of 3rd magnitude. The area is 249 sq deg. There is not much of telescopic interest in Indus. In this region these objects are somewhat clustered but nearly all of them are small and faint. This constellation contains the 5th magnitude star Epsilon, seven light-years distant and the nearest to the Earth of all naked-eye stars except Alpha Centaurus.

inelastic collision. A collision between two particles in which changes occur both in the internal energy of one or both of the particles and in the sums, before and after collision, of their kinetic energies. During collision, the "lost" kinetic energy is converted to internal vibration, noise, heat, etc.

Inertia, Law of. Newton's First Law of Motion which states: "Every body perserveres in its state of rest or of uniform motion in a straight line except in so far as it is made to change that state by external forces."

inertial coordinate system. A coordinate system not undergoing rotation or acceleration (with respect to the distant stars) in which Newton's first law is valid.

inferior conjunction. The conjunction of an inferior planet and the Sun when the planet is between the Earth and the Sun.

inferior planets. The planets with orbits smaller than that of Earth: Mercury and Venus. The inferior planets, Mercury and Venus, always remain near the Sun in the sky, whereas the outer, or superior planets circle the zodiac and can appear 180° from the sun. (See Fig., p. 116.)

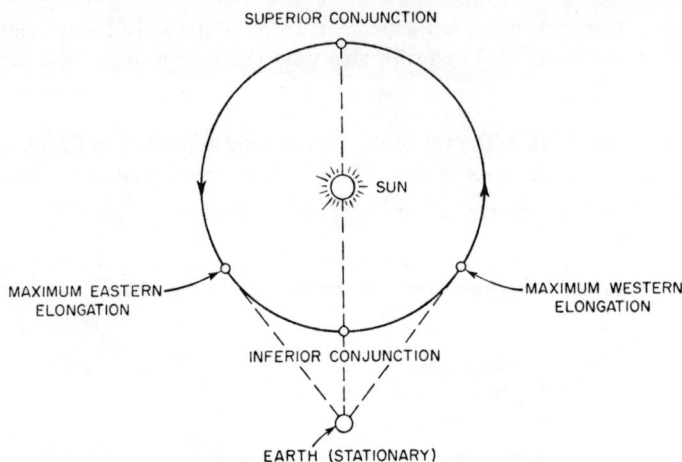

SUPERIOR CONJUNCTION

SUN

MAXIMUM EASTERN ELONGATION

MAXIMUM WESTERN ELONGATION

INFERIOR CONJUNCTION

EARTH (STATIONARY)

Phases of an inferior planet.

infinity.   An immeasurably large quantity, approached in the limit by $1/n$ as n approaches zero.

infrared radiation.   Radiation of wavelength longer than the limit which the eye can detect around 0.7 μn. Light from 0.7 to a few μns is said to be "near infrared," that beyond a few μns, "far infrared." Astronomer William Herschel, then living in England, first showed that invisible rays capable of increasing the temperature of a thermometer were present in sunlight just beyond the red end of the spectrum (Latin: *infra*, "below").

Inner Lagrangian point.   The point in a binary system where the Roche lobes meet. It is through this point that any transfer of mass from one component to the other takes place.

inner planets.   The four planets nearest the Sun: Mercury, Venus, Earth, and Mars. Sometimes used to indicate just the inferior planets, Mercury and Venus.

insolation.   In general, solar radiation received at the Earth's surface. The rate at which direct solar radiation is incident upon a unit horizontal surface at any point on or above the surface of the Earth.

intensity.   Energy received per unit area per second per unit solid angle, usually in the form of electromagnetic radiation but also as sound waves or particles such as cosmic rays.

interference filter.   A filter using interference of light waves to pass only certain wavelengths.

interference fringe (telescope). The dark lines obtained by passing two pieces of flat glass together, caused by alternate interference and reinforcement of light waves.

interferometer. An instrument that measures the small difference in direction of light or radio waves arriving from two sources by the interference of the waves. It can also be used to measure the direction of a radio source. The interferometer's operation depends on the wave properties of light. When two waves of light are out of step or out of phase, they "interfere" and cancel each other. When they are in step, they reinforce one another. In a plane where the relative phase of two waves varies, one observes a pattern of alternating dark and bright bands or fringes. A. A. Michelson, an American, first applied this property to the detection of small angular separation of double stars and the edges of stellar disks.

intergalactic space. Space between galaxies.

International Astronomical Union (I.A.U.). An organization coordinating the work of astronomers throughout the world. A number of committees have been established to specialize in various important departments of astronomy; these hold discussions at symposia held in various countries. I.A.U. conferences are held every three years.

International System of Units. A system of expressing measurements in the metric system by standardized units and abbreviations. The seven standard units are the meter, kilogram, second, ampere, kelvin, mole, and candela (measuring length, mass, time, electricity, temperature, chemical amount, and brightness, respectively). The system is often designated ISU, or by the French acronym SI. An important feature of the system is the set of prefixes indicating multipliers of the basic units, given in Table 1 of the Appendices.

interpolation. Estimation of the value of a variable quantity by using the trend of observed values bracketing the unobserved value in question. See also extrapolation.

interstellar dust. Microscopic particles distributed in the regions between the stars. They are detected by their effects of dimming and reddening the light of the distant stars. They are typically about $1/2\,\mu$m in size. Their composition is uncertain, possibly involving mixtures of silicates, ices, metals, and carbon compounds.

interstellar extinction. Dimming of starlight due to absorption by interstellar dust and gas. In the Milky Way plane it can amount to 1.9 magnitudes per kiloparsec of distance, but it varies from region to region.

interstellar lines. Absorption lines caused by interstellar gas appearing in the spectra of distant stars.

interstellar reddening. Reddening of the color of distant stars caused by the scattering of blue light out of the starlight beam by interstellar dust. Similar to the reddening of sunsets by dust in the Earth's atmosphere.

interstellar space.   Space between the stars.

intrinsic variable.   A variable star in which the observed variations in brightness are produced by some property inherent in the star itself (i.e., the pulsating and eruptive variables).

inverse square law.   The law which states that the intensity of a radiant phenomenon, such as light or gravity, varies inversely as the square of the distance from the source.

inversion layer.   An atmospheric layer in which temperature rises (instead of falls) with altitude.

inversion temperature.   The temperature at an altitude where the normal temperature lapse rate reverses and begins to rise (instead of fall) with increased elevation.

Io.   The 3rd largest satellite of Jupiter orbiting at a mean distance of 421,800 km. Its diameter is about 3500 km. Also called Jupiter I.

ion.   Charged particle formed when a neutral atom or molecule loses or gains one or more electrons.

ion column.   The trail of ionized gases in the trajectory of a meteoroid entering the upper atmosphere.

ionization.   (1) The process by which neutral atoms or groups of atoms become electrically charged (either positively or negatively) by the loss or gain of electrons, or the state of a substance whose atoms or groups of atoms have become thus charged. (2) The number (usually per cubic centimeter) of ions in a given region.

ionopause.   The surface between the ionosphere and the exosphere.

ionosphere.   A high-altitude region of the Earth's atmosphere containing a high concentration of ions including the D, E, F_1, and F layers that reflect or absorb radio waves. It is formed by the impact of solar x-rays and ultraviolet radiation on the atoms of the atmospheric gases. The D region occurs only in the daytime; the E region has a very low ionization density in the night; the F region persists throughout the night. The electron density varies in step with the 11-year sunspot cycle and the 27-day period of the Sun's rotation. Radio waves above about 30 cm/sec are not reflected by the ionosphere.

ionospheric storm.   Disturbance of the ionosphere, often by unusual solar radiation or particles, resulting in anomalous variations in its characteristics and effects on radio communications.

iron meteorite.   Any meteorite composed mainly of nickel–iron alloys, especially kamacite and taenite. They are silvery under their blackened exteriors. Where they occur in crystal forms, a characteristic pattern of intersecting crystal bands can be seen when the meteorite is etched with dilute nitric acid on a plane polished section. This Widmanstatten pattern is a test for meteorite origin of iron pieces.

irradiation. (1) Exposure to electromagnetic or particle radiation; (2) the apparent augmentation in size of a celestial body due to its brightness against the dark sky.

irregular galaxy. A galaxy of irregular shape. See galaxies, irregular.

irregular nebula. A diffuse and generally chaotic mass of luminous gas usually associated with interstellar dust.

irregular variable. A variable star for which no regular periodicity can be discovered in the light changes.

isentropic. Having constant entropy.

isobar. A line drawn on a map or chart through places having the same atmospheric pressure at a given time and at a standard level.

isoclinic line. A line through points on the Earth's surface having the same magnetic dip.

isodynamic line. A line connecting points on the Earth's surface which have the same total magnetic intensity.

isogenic lines. Lines on charts of the Earth's surface connecting points having equal magnetic declination.

isopleth. On a chart or graph, a line connecting points with constant value of a given quantity.

isostasy. The condition of hydrodynamic equilibrium in a planet's crust in which, for example, mountains are supported by low-density roots, like floating icebergs.

isotherm. A line or surface connecting points having the same temperature.

isothermal. A physical process which takes place without change in temperature. Opposed to adiabatic, where the process takes place without change in heat content.

isotopes. Two atoms of the same element with different masses because they have different numbers of neutrons. Unstable isotopes undergo radioactive decay. Atoms of stable isotopes normally remain unchanged.

isotropic. The same in all directions.

J

jansky. A unit of radio signal strength used by radio astronomers and equalling 10^{-26} watts per square meter per hertz. Named after a pioneer in radio astronomy, K. G. Jansky.

Janus. An inner satellite of Saturn with an orbital radius of about 160,000 km located on the outskirts of the ring system. The diameter is estimated at roughly 370 km.

Jeans criterion. A criterion for determining under what conditions an interstellar cloud or other massive object in a dispersed medium may become gravitationally unstable and begin to contract.

Jeans length. A characteristic dimension of a density disturbance (such as a sound wave) in a dispersed medium, larger than which the disturbance will become gravitationally unstable and begin to contract into a discrete mass. This condition is often invoked in studies of the formation of stars out of interstellar gas.

Johnson–Morgan photometric system. See photometry, three-color.

Joule's law. Heat is a form of energy and can be transformed into mechanical energy. James P. Joule showed this by experiments establishing that when mechanical energy is put into a system, the energy that "disappears" into the system reappears in the form of heat.

Jovian. (Latin: *Jovis*, genitive of Jupiter). Of or pertaining to the planet Jupiter.

Jovian planet. Any one of the giant planets: Jupiter, Saturn, Uranus, or Neptune. A giant planet.

Julian calendar. See calendar, Julian.

Julian date. A system of dating proposed by J. Scaliger in 1582 whereby a date is expressed as the number of days elapsed since the beginning of the arbitrary "Julian era" (January 1, 4713 B.C.). The Julian date is now approaching 2 1/2 million.

Juno. Asteroid number 3, the third discovered. Once thought to be one of the few largest asteroids, it is now believed to be surpassed in size by many later-discovered asteroids with low albedo. Its diameter is roughly 250 km. Its orbit lies in the asteroid belt with semimajor axis 2.7 A.U.

Jupiter. The largest planet, with mass about 308 times that of Earth and about 1/1000 that of the Sun. At 142,800 km, its diameter is about 11 times that of the Earth.

It is cloud-covered with an atmosphere believed to consist of about 60 percent by mass of hydrogen, 36 percent helium, a few percent of neon, water, ammonia, argon, and methane, and traces of other compounds that add rich colors to the clouds. Infrared measurements show that it radiates roughly twice as much energy as it receives from the Sun. The excess energy is believed to come from a slow gravitational Helmholtz contraction (q.v.) of its interior, a process perhaps continuing since Jupiter's formation. It rotates in about $9^h 50^m$ (equatorial cloud period). Its surface gravity is about 2.6 times that of Earth, and its escape velocity is about 61 km/sec. The surface nature is uncertain, though some lower atmosphere layers may approximate earth's temperature and pressure. The interior is believed to contain a small rocky core surrounded by liquid hydrogen, much of it in a high-pressure metallic form. Radio emissions and spacecraft measurements indicate a strong magnetic field several times stronger than Earth's and probably generated by currents in the metallic hydrogen interior. At least 14 satellites are known, including four large (Galilean) moon-sized satellites. The discovery rate of small outer moons in recent years suggests that more may be found.

Io.   In March 1979, Voyager I photographed orange and white surface markings believed to be sulfur deposits, and actively erupting volcanoes. The volcanic energy source is heat caused by flexing of Io by tides from massive Jupiter, making Io the most geologically active world known.

Europa.   Voyager photos showed long fissures or crack-like features on the surface, up to about 1000 km long and a few km wide.

Ganymede.   Voyager photos in 1979 showed cratered plains resembling lunar features, and also systems of unfamiliar bright streaks containing parallel grooves a few km wide; they may be fracture systems in the icy crust believed to cover Ganymede.

Callisto.   Voyager photos in 1979 revealed a heavily cratered surface including one vast probable impact basin 600 km wide and surrounded by a system of a dozen or more concentric rings as much as 2000 km across, resembling lunar multi-ring basin systems.

K

Kellner eyepiece. This positive eyepiece has a convex or plano-convex field lens; the eye lens is an achromatic plano-convex lens. The field of view is normally quite large and flat. A Kellner eyepiece is most useful for low powers when finding the field of a variable star.

k-corona. The inner portion of the solar corona, whose radiation is a continuous spectrum scattered by electrons.

kelvin. A temperature scale whose zero point lies at absolute zero ($-273°$ C). Same as absolute temperature. Abbreviated K, and not $°$K.

Kenneley–Heaviside layer. Early (1901–1930) name for Earth's ionosphere, especially the E layer, named after its discoverers.

Kepler's Harmonic Law. The law discovered by Johannes Kepler (1571–1630), that the square of the periods P of revolution of two planets are to each other as the cubes of their mean distance a from the Sun. Expressed as $P^2 = a^3$, when the period is in years and the distance is in astronomical units.

Kepler, Johannes. German mathematician and astronomer (1571–1630). Discovered the three laws of planetary motion showing that planets move around the Sun in elliptical orbits, thus establishing the idea of the solar system in its modern form.

Kepler's laws. The three empirical laws governing the motions of the planets in their orbits discovered by Johannes Kepler: (a) the orbits of the planets are ellipses, with the Sun at a common focus; (b) as a planet moves in its orbit, the line joining the planet and sun sweeps over equal areas in equal intervals of time (also called law of equal areas); (c) the squares of the periods of revolution of any two planets are proportional to the cubes of their mean distances from the Sun.

kilometer. A unit of length; 1,000 meters; 3,280.84 feet; approximately 0.62 miles. The number of kilometers in an interval approximately equals the number of miles times 1.6 (or more accurately, 1.60934).

kinetic energy. Energy of motion, $1/2\ mv^2$, equal to the work done in pushing a mass m until it moves at speed v, or the energy expended if such an object is stopped by a collision.

kinetic temperature. Temperature required to reproduce the velocity distribution observed in a gas.

kinetic theory of gases.   A highly successful mathematical description of gases as consisting of moving particles (molecules). It states that the average squared velocity of the molecules varies directly as the absolute temperature T of the gas and inversely as its mean molecular weight m. The average speed in km/sec at 0°C is 1.9 for hydrogen, 0.6 for water vapor, and 0.5 for nitrogen and oxygen. These speeds become 17 percent greater at 100°C. Each molecule of any gas has kinetic energy $1/2\ mv^2 = 3.2\ kT$, where k is the Boltzmann constant 1.38×10^{-16} erg/degree.

Kirchhoff's Laws (Kirchhoff-Draper Laws).   Three laws with many applications in astrophysics which govern the emissions and absorption of radiation by atoms under various physical conditions. They are often attributed to the German physicist G. P. Kirchhoff who published them in 1859 and sometimes also associated with the American astronomer W. Draper. (1) An incandescent solid, liquid or gas under high pressure radiates a continuous spectrum containing all colors or a wide range of colors. (2) An incandescent gas under pressure radiates a discontinuous or bright-line spectrum; that is, it emits light of a limited number of distinct colors characteristic of the particular element or compound. (3) When white light traverses a cool gas, the gas absorbs those colors that it would itself emit if it were incandescent. Such a gas will therefore produce an absorption, or dark-line spectrum.

Kirkwood Gaps.   Sharply defined regions of minimum number density of asteroids in the asteroid belt, seen as marked minima in a graph showing the number of known asteroid orbits versus semimajor axis (or revolution period). This is due to the cumulative effect of the peturbations caused by Jupiter when the period of the asteroid is a simple fraction such as 1/2 of the period of Jupiter.

Kleinmann-Low nebula (or KL nebula).   A massive infrared-emitting dust cloud inside the Orion nebula, perhaps the site of a newly forming open star cluster.

K line.   One of the Fraunhofer lines due to ionized calcium at λ 3933A. One of the most prominent lines in spectra of the Sun and solar-type stars.

knife edge.   Term used by mirror makers for any sharp edge used to cut a cone of light during optical tests. It is usually a razor blade. During common tests, the optician examines patterns of light reflected off the mirror as the knife edge moves across the beam near the eye.

Kramers' opacity.   A theoretical description of opacity in stars used in astrophysical analysis of light transmission in stars' interiors and atmospheres.

L

Lacerta (The Lizard). A small northern constellation half-immersed in the Milky Way between Cygnus and Andromeda extending to about 56° N. Hevelius published it first in 1687 as including ten stars. The area is 201 sq deg. There are seven open star clusters but only two of them make good telescopic objects. Three planetary nebulae have been found. The few extragalactic nebulae are small and faint.

lag. The failure of an indicating device to indicate immediately the variations in the property being measured.

Lagrangian points. In the plane of a circular-orbiting binary system five points where small bodies could theoretically co-rotate at the same period as the two larger bodies, thus keeping constant orientation with respect to the larger two bodies. In a system with one massive body M and a less massive body m, the only two stable points are located 60° ahead of and behind m in its orbit. In the Sun–Jupiter system these two points are actually populated by swarms of small bodies—the Trojan asteroids— which may have formed near these points or may have been captured into orbits located there.

Lambda Scorpii. A star of magnitude 1.6 which together with the nearest star Nu Scorpii marks the stinger on the tail of the Scorpion. It is at an estimated distance of 330 light-years, a B star, and a spectroscopic double.

Lambda Tauri. An Algol-type eclipsing binary with a range from 3.3 to 4.2 and a period of 3.9 days.

Lambert surface. An idealized diffuse reflector in which the amount of reflected light is proportional to the cosine of the emission angle (angle from ray to line-normal to surface).

lap. Any surface upon which a mirror may be polished. It must be yielding enough to embed the polishing agent in its surface. Common materials used are pitch, beeswax, and plastic.

lapse rate. The rate of decrease of temperature with elevation.

Large Magellanic Cloud (LMC). The larger of two irregular galaxies visible from the Southern hemisphere and believed to be satellites of the Milky Way. It is roughly 52 kpc (170,000 light-years) away and about 7 kpc (23,000 light-years) in diameter. The LMC is approximately circular, with its brightest portion or "axis" south of the center. This is a dense cloud of stars 1,000 by 5,000 light-years in extent, containing an even denser nucleus. Other bright regions in the LMC contain various kinds of clusters and nebulae, and highly luminous stars. The LMC also contains 30 Doradus, the largest and brightest of all known nebulae, described by some astronomers as a partially-formed nucleus for this galaxy.

large ion.   An atmospheric ion of relatively large mass and low mobility which is produced by the attachment of a small ion to an Aitken nucleus.

last quarter.   The phase of the Moon when it is near west quadrature so that the eastern half of it is visible in the morning sky to an observer on the Earth.

late-type stars.   The cooler stars, usually of spectral classes K and M. The term derives from an early but erroneous theory of stellar evolution.

latent heat of fusion.   The quantity of heat necessary to change 1 gram of a solid to a liquid with no change in temperature.

latent heat of vaporization.   The quantity of heat necessary to change 1 gram of a liquid to vapor with no change in temperature.

latitude, celestial.   The angular distance from the ecliptic, + measured to the north and − to the south.

lava.   Magma that has reached the surface of a planet and flowed out across it.

Law of Areas.   The line joining any planet to the Sun sweeps over equal areas in equal intervals of time. It is Kepler's Second Law of Planetary motion.

law of conservation of energy.   The energy in a closed system may be transformed into other forms, but the total amount remains the same. Energy can neither be created nor destroyed.

Laws of Motion.   Laws describing motions and accelerations of objects. See Newton's Laws.

laws of radiation.   See radiation laws, Kirchhoff's Laws.

law of reflection.   The angle of incidence is equal to the angle of reflection.

law of refraction.   A ray of light is bent towards the perpendicular to the surface when entering a denser medium and away from the perpendicular when entering a less dense medium.

law of spectroscopic analysis.   See Kirchhoff's Laws.

law of universal gravitation.   See gravitation and Gravitation, Law.

L-corona.   That portion of the radiation from the solar corona consisting of coronal line emission.

leap year.   The year divisible by four, to which an extra day is added in February to keep the civil calendar in step with the Sun. Century years (e.g., 1900) are not leap years, in order to make the calendar correction even more precise.

lemon-peel surface (telescope).   The surface of a mirror on which the imperfections resemble the skin of a lemon.

lens.   A transparent substance having two optically worked surfaces (at least one curved). Both surfaces are usually spherical, but may be convex (most common), concave, or planar.

lenticular.   Lens-shaped.

Leo (The Lion).   The 5th constellation of the zodiac, an ancient group to which Ptolemy assigned 27 stars. It is large and conspicuous in two groups, the first an upright sickle with the leader Regulus capping the handle, and the second an irregular trapezium of six stars. The area is 947 sq deg.

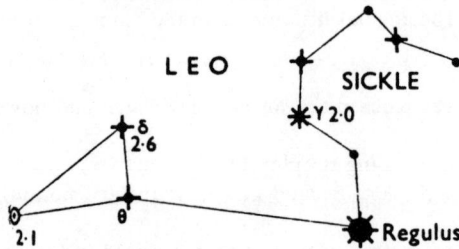

Leo Minor (The Small Lion).   An inconspicuous northern constellation formed by Hevelius about 1660 from 18 stars between Leo and Ursa Major. The area is 232 sq deg. It contains a few galaxies, but none bright enough to be of general interest. Three 4th magnitude stars and one of 5th magnitude in the form of a lozenge or diamond characterize this small constellation. Visible from January to July.

Leonids.   See meteor showers, Leonids.

leptons.   The lepton family have the lightest masses of all subatomic particles. They include the electron along with positrons, muons, and neutrinos. The positron is a positively charged elementary particle otherwise similar to an electron; it is the antiarticle of an electron. The muon is similar to the electron except that it is unstable and about two hundred times heavier. Neutrinos have no charge, no rest mass, and travel with the speed of light. Neutrinos are emitted during nuclear fusion on the Sun and carry some of the Sun's energy to the Earth. As the neutrinos are created in the Sun's core, they move right through the mass of the Sun as if it were empty and out into space, being little affected by intervening mass. Detection of neutrinos was accomplished in 1958 by two physicists, Clyde L. Cowan and Frederick Reines working with the nuclear reactor at the Atomic Energy Commission's Savannah River Plant. (see Fig., p. 127.)

The Leptons

Name	Symbol	Spin	Mass (million electron volts)	Mean Lifetime (seconds)
e-neutrino	ν_e	1/2	0	stable
μ-neutrino	ν_μ	1/2	0	stable
positron	e^+	1/2	0.5	stable
electron	e^-	1/2	0.5	stable
muon	μ^+	1/2	106	2×10^{-6}
	μ^-	1/2	106	2×10^{-6}

Lepus (The Hare). One of the ancient constellations immediately south of Orion. Ptolemy assigned it 12 stars. The 4 chief stars form a trapezium which permits easy recognition. The area is 290 sq deg. Lepus has no bright stars but Alpha (2.5) and Beta (2.8) are reasonably prominent.

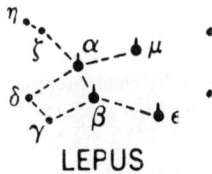

LEPUS

Libra (The Scale). The 7th zodiacal constellation received its name in Roman times, for it was originally the claws of Scorpius, the following constellation. Ptolemy assigned eight stars to these claws. The four principal stars, none brighter than the 3rd magnitude, form a trapezium preceding the head of Scorpius. The modern area is 538 sq deg.

LIBRA

libration. The axial swinging of the Moon with respect to the Earth due to its varying orbital velocity and other geometric and dynamical effects. Thus we can see the moon turned a bit E, W, N, or S of its normal position and thus glimpse portions of the normally averted hemisphere. See also libration, diurnal.

libration, diurnal (Moon).   A consequence of the Earth's rotation. Even if the other librations were absent (so that the same hemisphere were always turned exactly toward the center of the Earth) an observer on the eastern edge of the Earth (as seen from the Moon) would view the Moon from slightly different directions and therefore see slightly different hemispheres than an observer on the western edge. From the elevated position nearly 4,000 miles above the center of the Earth, the observer can see about 1° (lunocentric) farther over the western edge at moonrise and over the eastern edge at moonset.

light curve.   The curve representing the array of points where the observed magnitude (or brightness) of an object (such as a variable star) is plotted against the time of the observations. If the same variation is repeated periodically, the times of successive maximum and minimum brightness can be derived and predicted.

light-gathering power.   The ability of a telescope to collect and focus light. The amount of light gathered increases in direct proportion to the area of the telescope objective or the square of its diameter (neglecting minor losses in the optical parts).

light-microsecond.   The distance which light travels in one microsecond; equal to 327.8 yd or 299.8 m.

light time.   The elapsed time taken by electromagnetic radiation to travel from a celestial body to the observer and the time of observation.

light, velocity of.   The speed at which light travels, equaling 2.997925×10^8 m/sec; 186,283 mi/sec.

light-year.   A large unit of distance—about 9.46×10^{12} km or 6×10^{12} miles. It is the distance that light travels in one year (3.17×10^7 sec) at a speed of 3×10^5 km/sec.

limb.   The edge of the apparent disk of a celestial body, as of the Sun, the Moon, or a planet.

limb brightening.   A condition sometimes observed on celestial objects, in which the apparent brightness of the surface increases from the center toward the limb, due to increased scattering of light at certain wavelengths due to the increased air path of the line of sight there. Earth shows limb brightening.

limb darkening.   A condition sometimes observed on celestial objects, in which the brightness of the object decreases as one scans from the center toward the edges of limbs of the object approached. The Sun and Jupiter exhibit limb darkening, caused by absorption of light in their atmospheres; the path length through the atmosphere increases as one scans toward the limb.

limiting magnitude.   The faintest magnitude observable with a given instrument under given conditions.

line broadening. Increase in the width of spectral lines due to various causes such as high pressure gas in the source (pressure broadening), rotation of the source, turbulence in the source, etc.

line of apsides. The line connecting the two points of an orbit that are nearest and farthest from the center of attraction, such as the perigee and apogee of the Moon or the perihelion and aphelion of a planet; the major axis of any elliptical orbit extending indefinitely in both directions.

line of nodes. (1) The straight line connecting the two points of intersection of the orbit of a planet, asteroid, or comet, and the ecliptic. (2) The line of intersection of the planes of the orbits of any satellite and a reference plane, as in a satellite/planet or binary star system.

line of sight velocity. The velocity of approach or recession between two bodies; its magnitude is determined by the Doppler effect. It may be less than the total relative velocity of the two bodies, since there may be a component of motion perpendicular to the line of sight between the two bodies. Also called radial velocity.

line profile. A tracing of a spectral line giving intensity as a function of wavelength.

lines. Deficiencies or excesses of light of certain colors (wavelengths) in the spectra of radiating bodies. When the spectrum is presented as a band of colors from blue to red (increasing wavelength), spectral lines appear as dark gaps (absorption lines) or sharp increases (emission lines) in the amount of light of certain colors. All can be designated by their wavelength.

lithosphere. The solid-rock part of the Earth or other planetary body. Distinguished from the atmosphere, hydrosphere, and molten or metallic core.

local apparent time. Time based on the position of the true sun in the sky. Because the Sun appears to move at a varying rate during the year, this type of time is not used in day to day life. The local hour angle of the apparent or true sun is expressed in time units, plus 12 hours. See also local mean time.

Local Group. About 20 of the nearest galaxies (including the Milky Way Galaxy) appear to form a cluster. Most of the mass is contained in the Milky Way and Andromeda galaxies.

local hour angle of a star. See hour angle.

local mean time (LMT). Time based on the position of an imaginary Sun called the "mean sun," which moves along the equator at the average rate of the true Sun. See also local apparent time.

local meridian. The meridian that happens to pass through one's zenith at any instant of time.

local sidereal time (LST).   Local hour angle of the vernal equinox expressed in time units; the right ascension of a star that is on the local meridian. Clocks keeping sidereal time are thus used in pointing telescopes at celestial objects of known position.

local standard of rest (LSR).   A coordinate system moving with the mean velocity of the local stars near the Sun (out to about 50 pc). Stars moving with local group of stars.

long period variable.   A red star in which the light changes are relatively slow and show a certain degree of regularity. The periods usually last between 100 and 1000 days, the majority being about 300 days. The amplitudes are usually large for a variable star, reaching 9 magnitudes in visual light. Mira is an example.

local thermodynamic equilibrium.   See LTE.

longitude.   The set of coordinates defined by equally spaced great circles through the poles of rotation of a planet or star. Longitude is usually specified by the number of degrees measured along the equator east or west of a reference meridian—one of the great circles.

look-back time.   The amount of time since light received from an object was emitted by the object. Look-back time in years equals distance in light-years.

looming.   A mirage effect produced by greater than normal refraction in the lower atmosphere, thus permitting objects to be seen that are beyond the geometric horizon. This occurs when the air density decreases more rapidly with height than in normal atmosphere.

low.   A pressure system characterized by relatively low pressure at the center.

lower atmosphere.   Generally (and quite loosely) that part of a planetary atmosphere in which most weather phenomena occur (i.e., the troposphere and lower stratosphere; used in contrast to the common meaning for the upper atmosphere.

lower branch.   That half of a meridian or celestial meridian from pole to pole which passes through the antipode or nadir of a place.

lower limb.   That half of the edge of the apparent disk of a celestial body having the least altitude above the horizon; in contrast with the upper limb, that half having the greatest altitude.

LTE.   Local thermodynamic equilibrium, the assumption that molecules in a certain region have velocities and other properties characteristic of equilibrium at a certain fixed temperature. The approximation is widely used in the construction of theoretical models of gases in stars and other atmospheres.

luminance.   In photometery, a measure of the amount of light reflected or emitted by a source in a given direction; luminous energy per second per unit solid angle per unit of projected arc as seen by an observer not necessarily looking perpendicular to the surface. It is weighted according to the sensitivity of the human eye, thus counting only the energy of visible wavelengths.

luminosity.   The actual or intrinsic brightness of a star. The total amount of energy radiated per second by the star. Usually expressed as ergs/sec. The Sun's luminosity is about 4×10^{33} erg/sec.

luminosity classification.   W. W. Morgan and P. C. Keenan at Yerkes Observatory established in the 1940's the following luminosity classes:

Ia   Most luminous supergiants
Ib   Less luminous supergiants
II   Bright giants
III Normal giants
IV   Subgiants
V   Dwarfs or "main sequence."

These classes can be distinguished in most cases from spectra of the stars.

luminosity function.   The number of stars or other objects per unit volume of space for various luminosities.

luminous.   Emitting visible radiation.

luminous efficiency.   For a given wavelength of radiation, the ratio of the energy that is effectively sensed by the human eye to the energy that is intrinsic in the radiation. It may be represented as a dimensionless ratio (i.e., lumens per watt). It is thus high for yellow light and low for far red or blue.

luminous emittance.   The emittance of visible radiation weighted to take into account the different response of the human to different wavelengths of light.

luminous energy.   The energy of radiation weighted in accordance with the wavelength dependence of the response of the human eye.

luminous flux.   Luminous energy per unit time; the flux of visible radiation, so weighted as to account for the manner in which the response of the human eye varies with the wavelength of radiation.

lunar tide.   The tide due to the gravitational attraction of the Moon alone, neglecting the sun.

lunar calendar.   See calendar, lunar.

lunar cycle.   Any cycle related to the Moon, particularly the Callippic cycle or the Metonic cycle.

lunar day. (1) The duration of one rotation of the Earth on its axis, with respect to the Moon. Its average length is about 24 hours 50 minutes of mean solar time. Also called tidal day. (2) The rotation period of the Moon with respect to the Sun, or synodical month equals 29.53 days—the period of nighttime and daytime at a fixed lunar site.

lunar eclipse. The phenomenon observed when the Moon enters the shadow of the Earth. An eclipse of the Moon. Below is the configuration of the Sun (S), Earth (E), and Moon (M) during a lunar eclipse (not to scale). See also eclipse.

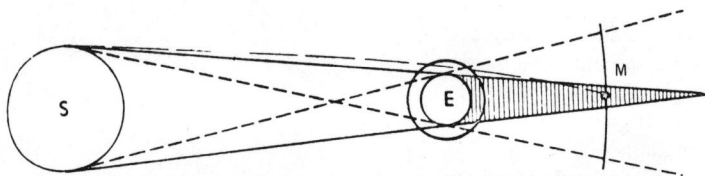

lunar features (basins). The largest circular features, often with multiple concentric ring-shaped rims and often filled with mare lavas. Not originally recognized as craters by telescopic observers, but now believed to be the largest examples of impact craters.

lunar features (craters). Circular depressed structures with raised rims believed to be caused mostly by meteorite impacts, though some may be volcanic. The largest were once believed to be 100 to 200 km in diameter, but recent studies indicate that large mare-filled basins up to 1000 km are also impact craters. The smaller the crater, the more common it is. The smallest are microscopic pits that dot the surfaces of exposed rocks. Ejected material blasted out of craters form bright rays (or radial streaks of bright material) and secondary impact craters (formed by ejected debris hitting the surface) up to a few percent of the size of the parent primary crater.

lunar features (domes). These curious features, like low swellings or hillocks, are of heights up to a few hundred meters and diameters of several kilometers; they can be clearly seen only when near the terminator; more than a hundred are known. They may be volcanic features.

lunar features (Maria or seas). These form the lunar lowlands and are dark, flat plains composed of basaltic lava flows. The Latin name for "seas" (singular, *mare*) was bestowed on them by the early telescopic observers, who thought they really were liquid bodies. The maria differ in appearance from the highlands in being both darker and smoother. Winding ridges are common, and so are small hillocks and craterlets, but they have accumulated few prominent craters because they are younger than the uplands. Typical ages determined from lunar rocks are 3.2 to 3.8 billion years. Here and there are low rings, often of considerable size; these "ghosts" (so-called because of their faintness) are evidently the walls of ancient craters that were overlaid by molten lava when the seas were formed.

lunar features (names).   The system of crater-naming follows the plan suggested in 1651 by the Italian observer Giovanni Riccioli who called them after famous scientists and other prominent figures. The Moon therefore houses such personages as Plato, Newton, Archimedes, and Copernicus. Riccioli himself supplied about 200 names for the craters he plotted, but subsequent work has added more than a thousand listed formations. Twentieth-century scientists and others have been added as names on the Moon's far side.

lunar gravity.   The acceleration imparted by the Moon to a mass which is at rest on the Moon. It is approximately 1/6 of the Earth's gravity, or 162.2 cm/sec^2. Thus objects weigh about 1/6 as much on the Moon as they do on the Earth.

lunar month or lunation.   The interval between successive times of new moon. Sometimes called the synodical month, 29.53 days.

lunar space.   That part of space in which the gravitational attraction of the Moon predominates.

lunation.   The period elapsing between successive similar phases of the Moon, also known as the synodic period. It is roughly 29.5 days.

lunisolar calendar.   This complex calendar tries to keep in step with the Moon's phases and the seasons. It began by occasionally adding a 13th month to the short lunar year to round out the year of the seasons. The extra month was later inserted by fixed rules. The Jewish calendar is the principal survivor of this type.

lunisolar precession.   As a consequence of its equatorial bulge, the spinning Earth behaves like a gyroscope. The net result is a precession or wobbling of the position of this axis. Earth axis thus retains the same angle of 23.5 degrees to the ecliptic poles, but continuously changes its direction in space. The axis will traverse the surface of a cone once in approximately 26,000 years in a direction opposite to Earth's rotation. This is the lunisolar precession—the precession due to the combined action of the Sun and the Moon. Below is a diagram of the lunisolar precessional motion of Earth's axis. This precession causes a continuous slow change in the right ascension and declination coordinates of all stars, since those coordinates are defined by the position of Earth's rotation axis. (See Fig., p. 134.)

Lupus (The Wolf).   Designated in the Almagest of Ptolemy as the wild beast, to which 19 stars are assigned. It lies mainly between Centaurus and Scorpius, partly immersed in the Milky Way, and near the region in which globular clusters are scattered so profusely. The modern area is 334 sq deg and includes many stars of moderate brightness without definite pattern. Lupus lies in a rich region of the sky and has many attractive telescopic objects including fine doubles and planetary nebulae. Three globular clusters and four open clusters are known. The few galaxies in this heavily veiled region are faint and small.

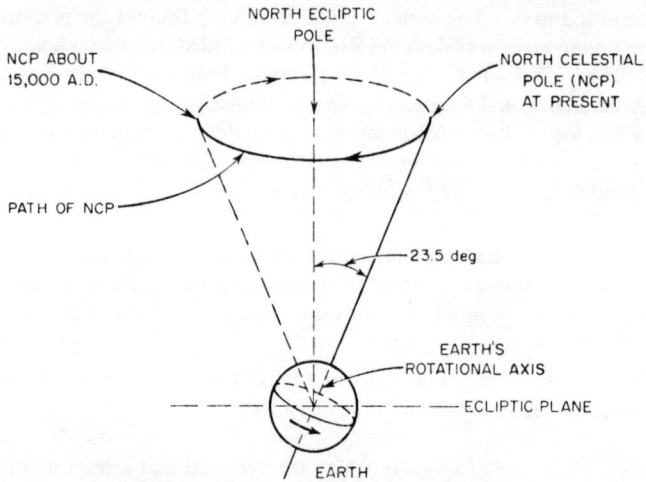

Lyman alpha.   A line in the far-ultraviolet part of the spectrum (1215 Å) strongly absorbed and easily emitted by hydrogen atoms. Lyman beta, Lyman gamma, Lyman continuum, and others are similar but at shorter wavelengths.

Lyman series.   The series of spectral lines produced in hydrogen atoms by electrons transferring to or from the first energy level (or orbit) closest to the nucleus.

Lynx (The Lynx).   A fairly large constellation charted by Hevelius (about 1660) from 19 stars. It lies in a very barren area of the sky between Auriga and Ursa Major with no prominent stars to mark it. The area is 545 sq deg. There is one globular cluster NGC 2419 which is one of the most distant galaxies. Most of the numerous galaxies are small and faint, the brightest being NGC 2683.

Lyot telescope.   A telescope for studying solar prominences by means of a metal disk located at the focal plane in the tube of the telescope.

Lyra (The Lyre).   A very ancient constellation which received its present name from a Homeric myth. Ptolemy assigned 10 stars to it and it may be recognized by the compact group of stars between Hercules and Cygnus, led by the brilliant Vega, the 4th brightest star in the sky. The area is 286 sq deg. There are some fine double stars in Lyra. Two planetary nebulae are known including the well-known ring nebula NGC 6720.

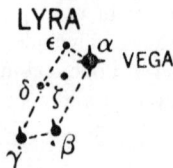

Lyrids.   A now weak meteor shower occurring annually about April 21. It has been traced back to 15 B.C. and is associated with the comet of 1861. The Lyrids have an orbit inclined 80° to the ecliptic.

M

mackerel sky. A popular name for cirrocumulus clouds.

macro. A prefix referring to dimensions ranging from greater than microscopic to cosmic.

Maffai 1 and 2. Two nearby galaxies obscured by interstellar dust in our own galaxies. They were discovered in 1970 by infrared photography which penetrates the intervening dust.

magellanic clouds. In the southern hemisphere, the naked eye can easily make out two very nearby galaxies: the Large Magellanic Cloud (apparent angular diameter about 7°) and the Small Magellanic Cloud (about 2 1/2°). These are believed to be satellite galaxies orbiting around the Milky Way. The telescopic photograph of the two Magellanic systems reveals that their structure is quite different from that of our own galaxy or our neighboring spiral, M31. They are irregular galaxies—amorphous masses of stars about 60 kpc away from us. Their angular sizes translate into linear diameters of 7 kpc and 2.5 kpc for the large and small systems respectively. From the existence of so many O and B associations in both Magellanic systems, it is estimated that a significant portion of their material is still in the form of gas and dust, and that the formation of new stars is proceeding vigorously at this time. See also Large Magellanic Cloud and Small Magellanic Cloud.

magma. Molten rock underground. See also lava.

magnesium fluoride. A material which when coated 1/4 of a wavelength thick on a lens increases the amount of light transmitted and decreases the amount reflected.

magnetic declination. In terrestrial magnetism, the angle between true north and magnetic north at any given location.

magnetic dip. The angle between the horizontal and the direction of a line of force of the Earth's magnetic field at any point.

magnetic equator. That line on the surface of the Earth connecting all points at which the magnetic dip is zero.

magnetic field. A region in which a magnetic pole, compass needle, a moving electric charge, or a conductor carrying a current, is subjected to the action of a force called a magnetic force. Earth is surrounded by such a magnetic field (often called the geomagnetic field or the terrestrial magnetic field) originating in its interior, and having a strength about 1/2 gauss. In any magnetic field a compass needle aligns itself in the "direction" of the field, or along the so-called field lines.

magnetic lunar daily variation.   A periodic variation of the Earth's magnetic field that is in phase with the transit of the Moon.

magnetic meridian.   The horizontal line which is oriented at any specified point on the Earth's surface along the direction of the horizontal component of the Earth's magnetic field at that point, thus running from magnetic north to magnetic south.

magnetic north.   The direction indicated by the north-seeking needle of a magnetic compass; it is usually a few degrees different from true north, which is defined by the planet's rotation axis.

magnetic poles.   (1) The poles of opposite magnetic attraction at the two ends of a magnetic dipole. (2) The two locations representing the poles of unlike magnetism belonging to the Earth or other planets. On Earth, the North Magnetic Pole is currently located at approximately 73° N latitude, and 100° W longitude in Prince of Wales Island. The South Magnetic Pole is currently located at approximately 71° S latitude and 149° N longitude in Antarctica. The locations slowly change over the centuries.

magnetic stars.   Stars, mostly of spectral type Λ, with very strong magnetic fields ranging up to about 30,000 gauss.

magnetic storm.   Disturbance of the terrestrial magnetic field caused by a stream of solar particles traveling at about 6 million km/hr through interplanetary space, covering the distance from the Sun to the Earth in about one day. The storms are caused by particles shot out of solar flares and sunspots; violent ones have been correlated with the sunspot cycle and milder ones with the synodic period of solar rotation (27 days); the former last for hours and the latter for days.

magnetic variable.   A star in which the light changes are also accompanied by variations in the strength of the magnetic field.

magnetic variation.   The angle between geographic and magnetic north at any given place. Expressed in degrees east or west, to indicate departure of magnetic north from true north. Often simply called variation.

magnetograph.   An instrument for mapping the small magnetic field of the Sun utilizing the Zeeman effect. It scans the Sun and enables separations of the spectral lines themselves to be measured accurately with the aid of special filters. The trace of the measurements wander above and below the scanning line according to the sign and strength of the magnetic field at each point along the line.

magnetohydrodynamics.   Study of the motions of ionized gases in the presence of a magnetic field, in which case magnetic forces may completely dominate over gravitational forces.

magnetopause. The region above 60,000 km where the earth's atmosphere and magnetic field interact with the solar wind.

magnetosphere. The region of the Earth's atmosphere where ionized gas plays an important part in the dynamics of the atmosphere and thus where the geomagnetic field plays an important role in gas motions. The magnetosphere begins by convention at the maximum of the F layer at about 350 m and extends to 10 or 15 earth radii to the boundary between the atmosphere and the interplanetary plasma.

magnification. Referred to as a telescope's "power." The total magnification of a telescope is given by the focal length of objective divided by the focal length of the eyepiece. "Power 50X" means that an object (like the Moon) appears 50 times larger in the eyepiece than to the unaided eye.

magnifying power. See magnification.

magnitude. An astronomical term for indicating brightness, not size. Stellar magnitude can be thought of as the logarithm (on a special base) of the faintness. The larger the numerical magnitude, the fainter the star. The difference between the apparent magnitude of a star (how bright it looks to us) and the absolute magnitude (how bright it would look at 10 parsecs distance) becomes important in the measurement of distance. If one object is 5 magnitudes brighter than another, it is 100 times brighter. A brightness difference of 1 magnitude is a brightness factor of roughly 2 1/2 (2.512). The magnitude scale for visible stars is:

0   Extremely bright stars such as Rigel in Orion and Vega in Lyra.
1   Very bright stars, standing out among their neighbors. Conventionally, any star of magnitude brighter than 1 1/2 is said to be of the 1st magnitude; there are only 21 such stars in the whole sky.
2   Moderately bright stars, such as Polaris and the six senior members of the Great Bear.
3   Faint stars, but still visible when there is some mist or city light.
4   Fainter still, concealed by mist, moonlight, or city lights.
5   Too faint to be seen when the sky is not really dark and clear.
6   The faintest stars visible with the naked eye under good conditions.

The scale can be extended in either direction (negative numbers or higher positive numbers) for brighter or fainter objects. Brightness differences are shown in the following table:

Magnitude Difference	Brightness Ratio	Magnitude Difference	Brightness Ratio
1.0	2.51	7.0	631.0
1.5	4.0	7.5	1,000.0
2.0	6.3	8.0	1,585
2.5	10.0	8.5	2,512
3.0	15.9	9.0	3,981

Magnitude Difference	Brightness Ratio	Magnitude Difference	Brightness Ratio
3.5	25.1	9.5	6,310
4.0	39.8	10.0	10,000
4.5	63.1	11.0	25,120
5.0	100.0	12.0	63,096
5.5	158.5	13.0	158,490
6.0	251.2	14.0	398,110
6.5	398.1	15.0	1,000,000

magnitude, photographic. A measure of the brightness of a star or other celestial body in wavelengths to which a photographic film or plate is sensitive, usually blue-colored light. (See magnitude.)

magnitude, visual.   A measure of the apparent brightness of a star or other celestial body in wavelengths to which the eye is sensitive. (See magnitude.)

main sequence.   The stars that are consuming hydrogen as the source of their radiant energy. A large proportion, at least 90 percent, of the nearby stars in a Hertzsprung–Russell (H–R) diagram fall into a diagonal band called the main sequence. Main-sequence stars range from hot, massive O-type stars through G-type stars like the Sun, to cool, small, faint red stars. Other star types such as red giants, supergiants, variables, and white dwarfs are not main-sequence types. Below is a simplified representation of the Hertzsprung–Russell diagram showing the characteristics of the main sequence. (See Fig., p. 140.)

major axis.   The longest axis of an ellipse, upon which the foci are situated.

major planets.   The four largest planets: Jupiter, Saturn, Uranus, and Neptune.

Maksutov telescope.   A reflecting telescope using a spheroidal instead of a paraboloidal main mirror. Before reaching the mirror, the light passes through a concave lens almost as large as the mirror itself to remove the spherical abereration. A compound telescope that uses three optical elements: a correction lens, a spherical primary mirror, and a hyperboloidal secondary.

mamato-cumulus.   A type of cumulus cloud in which the lower surface bulges downward in a number of places to produce rounded protuberances. These clouds are often precursors of storms.

mantissa.   The part of a logarithm right of the decimal point.

mare.   (Pronounced mah′-ray; Latin for sea) (1) The large, dark, flat areas on the lunar surface that were thought by early astronomers to be bodies of water. (2) Dark areas on Mars and Mercury. Plural is maria (seas).

TEMPERATURE (°K)

25,000   11,000   7600   6000   4000

-5

10,000

SUPERGIANTS

1000

SPICA

ARCTURUS

0

100

SIRIUS

MAIN SEQUENCE

RED GIANTS
(POP I)

10

ABSOLUTE VISUAL MAGNITUDE

+5

1

SUN

0.1

+10

0.01

0.001

WHITE DWARFS

LUMINOSITY (RELATIVE TO SUN)

+15

0.0001

O   BO   AO   FO   GO   K   M

SPECTRAL CLASS

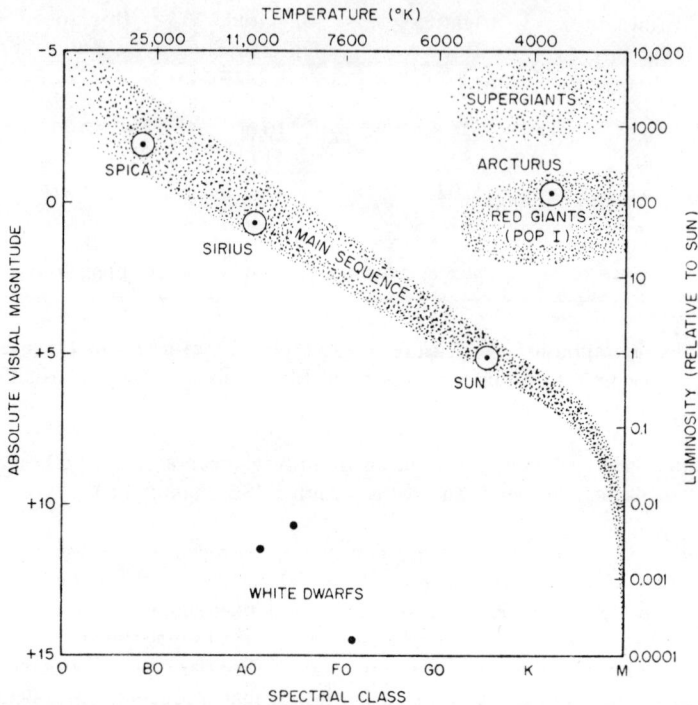

mare's tails.   A feathery, spreading cirrus cloud.

Markab.   See Alpha Persei.

Mars.   The 4th planet, the 1st beyond the Earth. It moves around the Sun 1.88 years at a mean distance of 1.52 A.U. (228 × 10⁶ km). The orbit is noticeably eccentric (0.093). The distance from Mars to the Sun varies from a minimum of 208 million km at perihelion to a maximum of 249 million km at aphelion. The apparent diameter of Mars varies from a maximum of 25″ at perihelic opposition when it is nearest to Earth, to 14″ at aphelic opposition, and to a minimum of 4″ when in conjunction with the Sun. Observations are best when the planet is within a few months of opposition. Mars has two small satellites, Phobos and Deimos, discovered in 1877 by A. Hall. Spacecraft have shown these to be cratered rocky chunks about 20 × 28 km and 10 km × 16 km, respectively. Spacecraft data show that one hemisphere of Mars is dominated by an ancient moon-like surface with many eroded craters. The other hemisphere contains many younger, less cratered volcanic plains, canyons apparently formed from geological faults, and several enormous volcanic mountains larger than any others known in the solar system. Two Viking landers in 1976 showed that the soil in two plains regions was of basalt-like composition and contained no organic materials (to an accuracy of a few parts per billion). No life forms were detected, though the soil showed unusual chemical properties, probably due to unusual mineralogical states produced by solar ultraviolet radiation that is not stopped by the thin Martian atmosphere. The atmospheric pressure averages around 0.7% that on Earth. Nighttime low temperatures of about $187°K$ ($-124°F$) were recorded at the two landing sites. The warmest daytime temperatures are believed to exceed freezing by several degrees on summer afternoons. The atmosphere composition by volume is approximately: 95 percent CO_2, 2.7 percent N_2, 1.6 percent Ar, 0.6 percent CO, 0.15 percent O_2, and traces of other gases including variable traces of H_2O. Polar ice deposits probably include both H_2O and CO_2. Today most of the H_2O is frozen in the polar caps or under the soil, or chemically bound in rock minerals. However, evidence of erosion (including dry channels resembling riverbeds and tributaries) has led many analysts to conclude that Mars may have had a warmer, more water-rich climate in the past. Photographic evidence from space probes indicates that the once-reported "canals" (q.v.) are mostly illusory, and that the dark patchy markings once suspected to be vegetation are formed by deposits of wind-blown dust.

Mars crossing asteroids.   See Asteroids, Amor.

Mars, oppositions.   Recur at intervals of the synodic period, which averages 780 days, or about 50 days longer than 2 years. Thus they come in alternate years and each time about 50 days later than before. Favorable oppositions occur at intervals of 15 to 17 years and usually in August or September because on August 28 the Earth has the same heliocentric longitude as the perihelion of Mars.

Year	Opposition	Nearest Earth	Millions of km	Diameter	Magnitude
1971	Aug 10	Aug 12	56.3	24."9	−2.6
1973	Oct 24	Oct 17	65.2	21.4	−2.2
1975	Dec 15	Dec 9	84.5	16.5	−1.5
1978	Jan 22	Jan 19	97.7	14.3	−1.0
1980	Feb 25	Feb 26	100.7	13.8	−0.9
1982	Mar 31	Apr 5	95.1	14.7	−1.1
1984	May 11	May 19	79.7	17.6	−1.6
1986	Jul 10	Jul 16	60.5	23.2	−2.4
1988	Sep 28	Sep 22	58.9	23.8	−2.5
1990	Nov 27	Nov 20	77.4	18.1	−1.7
1993	Jan 8	Jan 3	93.7	14.9	−1.2
1995	Feb 12	Feb 11	101.2	13.8	−0.9
1997	Mar 17	Mar 20	98.6	14.2	−1.0
1999	Apr 29	May 1	86.6	16.2	−1.4

maser.   A device (or sometimes a nebular interstellar cloud) in which atoms or molecules are excited to a condition with many electrons in metastable energy states. Incoming radiation of a certain wavelength can then cause electrons to return to a lower state in these atoms or molecules, releasing many photons of a specific wavelength in phase with each other. The resulting radiation in a given direction will be much stronger than the light radiated by thermal equilibrium processes alone. The name is an acronym for Microwave Amplification by Stimulated Emission of Radiation.

mass.   The amount of material in a body; it may be measured by the gravitational attraction of the body toward another body. It is also a measure of resistance to change of velocity, and varies as velocities increase, especially near the speed of light. Rest mass is the mass measured at zero velocity.

mass function.   (1) For single-line spectroscopic binary stars, the ratio of the cube of the product of the mass of the second body times the sine of the inclination of the orbit to the sum of the masses squared. (2) The relation between stellar mass and the number of stars/psc^3 of each mass.

mass–luminosity law.   The relationship that exists between the mass and the luminosity of a star, discovered by A. S. Eddington in 1924.

mass–radius relation.   A relation between mass and radius holding for main-sequence stars.

matter.   The universe can be divided logically into two categories: matter and energy. Matter is anything that occupies space and has mass (more specifically, rest mass). Matter exists in different forms including gas, liquid, solid, and high-density degenerate forms. It has been estimated that nearly all the material in the universe is in the gaseous form and that only one part of it in 1,000 is nongaseous.

Maxwellian distribution.   The number of atoms or molecules moving within different velocity ranges in a gas of specified temperature. Later generalized by Boltzmann in a form called the Maxwell–Boltzmann distribution.

mean.   Average. The sum of the measured values divided by the number of values measured.

mean free path.   Average distance a particle travels before colliding with a neighboring particle.

mean solar day.   The day determined by successive meridian crossings of the mean Sun. (A fictitious Sun moving along the equator at the average rate of the real Sun.)

mean solar time.   Time measured by the daily motion of a fictitious body called the "mean sun." Since the apparent sun travels along the ecliptic with a motion that varies during the year, it cannot be used to measure time.

mean sun.   An imaginary version of the Sun which travels through the sky at a constant angular rate without the variations (due to earth's noncircular orbit, etc.) that characterize the true sun (q.v.).

mega.   A prefix indicating 10^6, or one million. Abbreviated M.

megacycle.   One million cycles, a unit of radio frequency abbreviated Mc/sec or MH_z (for megahertz).

megaparsec.   A million parsecs, or 3×10^{24} cm.

meniscus lens.   A "bent" lens, or one with surfaces so arranged that its cross section is crescent-shaped.

Mensa (The Table).   Named after Table Mountain in South Africa, this constellation was introduced by Lacaille about 1752. It lies far south, between Dorado and the polar Octans, and its stars are inconspicuous. The area is 153 sq deg. Since Mensa has no star even as bright as the 5th magnitude, it would be of little note but for the presence of the Large Magellanic Cloud.

mercator projection.   A conformal projection on which the meridians and parallels are shown as parallel straight lines at right angles to one another, the divisions of latitude being expanded north and south of the equator in the same proportion as the divisions of longitude have been lengthened by projection. On this projection a line of constant bearing (or rhumb line) is represented as a straight line. High-latitude land masses appear enlarged in area relative to low-latitude land masses.

Mercury.   The mean distance of Mercury from the Sun is about 58 million km. Its orbit is inclined some 7° to the plane of the ecliptic and has a large eccentricity of 0.206. Because of this eccentricity the distance of Mercury to the Sun varies from a maximum of 46 million km at perihelion to a maximum of 70 million km at aphelion. Its orbital velocity varies accordingly from 39 km/sec at aphelion to 57 km/sec, the fastest of all planets. Spacecraft have shown that the surface of Mercury is quite lunar-like, being mostly covered with densely crowded craters. Some relatively smooth plains resemble the lunar maria. Features distinct from those on the moon include some massive cliffs which appear to be caused by compressive thrust faults, or fractures. Also, there are more areas of so-called inter-crater plains, which may be ancient lava flow regions, than on the moon. There is virtually no atmosphere. A magnetic field about 1 percent as strong as the Earth's, together with the planet's high mean density of 5.4 g/cm³, suggests that Mercury may have an iron core similar to Earth's core, in which the magnetism is generated.

meridian.   (1) The great circle of the celestial sphere which passes through the celestial poles and the observer's zenith. (2) A north–south reference line on a planet's surface passing through the geographical poles of rotation.

meridian circle.   A telescope constrained so that it can be pointed only at the meridian. It is used to time the instant at which stars pass across the meridian and to keep a check on the rotation of the Earth.

mesopause. (1) In the Earth's atmosphere, the upper limit of the mesosphere (temperatures roughly 180°K) usually found at 80 to 85 km. Above this level the temperature increases with height. (2) A similar region in other planets' atmospheres.

mesosphere. (Greek: *mesos*, "intermediate"). (1) The portion of the Earth's atmosphere from about 30 to 80 km in which the temperature, with increasing height, at first increases to a maximum around 280°K (45°F) and then decreases to about 180°K (−135°F) or less, depending on the latitude and the season. Its lower boundary is the stratopause and the upper mesopause. (2) A similar region in other planets' atmospheres.

Messier Catalog. 1781 catalogue of 103 nebulae clusters and galaxies designed by Charles Messier to aid comet hunters; objects in the catalogue are designated by M and a running number, such as M31.

Messier Number. In objects, numerical designation in the Messier Catalog, such as M31.

Messier Objects. The group of 103 nebulae and star clusters listed in Messier catalog.

metagalaxy. The measurable material universe of galaxies. The regions relatively near the Milky Way are called the inner metagalaxy, after a book title by Harlow Shapley.

metastable state. An excited state (upper energy level) in an atom or molecule, in which an electron may remain for a longer duration than in most other excited states. On Earth electrons are rapidly knocked out of these states by collisions between atoms or molecules, but in interstellar space collisions are rare, and many atoms and molecules may reach metastable states.

meteor. A celestial phenomenon associated with a streak of light of short duration; it is commonly known as a "shooting star." The bright streak is usually a microscopic or millimeter scale of matter traveling at high velocity into Earth's atmosphere, causing it to be heated to incandescence by air friction. The particle itself is sometimes called a meteor, but it is preferably designated as a meteoroid. The luminous phenomena typically occur between 130 and 40 km above the ground. A very bright meteor is known as a fireball, and a large fireball, particularly one accompanied by sparks and explosive noise, is called a bolide.

meteor showers. A cloud of small interplanetary meteoric particles in space encountering the Earth and causing unusually large numbers of meteors for several nights. Because the meteors move through the atmosphere in parallel paths (in 3-dimensional space), they appear to the observer to stream from a common point in the sky—the radiant. The radiant represents the direction from which meteors approach the Earth. Showers occur as the Earth intersects orbits of many known comets indicating that the meteoroid grains were ejected from the cometary nucleus and then scattered along the orbit. Because their velocities of ejection were small, they moved slowly away from the nucleus and are now moving around the Sun in individual orbits that do not differ much from that of their parent comet.

Most Prominent Nighttime Meteor Showers

Name of Shower	Period of Detectable Meteors	Date of Peak Activity	Maximum Visual Hourly Rates
Quadrantids	Jan 1–4	Jan 3	50
Corona Australids	Mar 14–18	Mar 16	(5)
Virginids	Mar 5–Apr 2	Mar 20	(less than 5)
Lyrids	Apr 19–24	Apr 21	5
Eta Aquarids	Apr 12–May 12	May 4	12
Ophiuchids	Jun 17–26	Jun 20	(20)
Capricornids	Jul 10–Aug 5	Jul 25	(20)
Southern Delta Aquarids	July 21–Aug 15	Jul 30	20
Northern Delta Aquarids	Jul 15–Aug 18	Jul 29	10
Pisces Australids	Jul 15–Aug 20	Jul 30	(20)
Alpha Capricornids	Jul 15–Aug 20	Aug 1	5
Southern Iota Aquarids	Jul 15–Aug 25	Aug 5	(10)
Northern Iota Aquarids	Jul 15–Aug 25	Aug 5	(10)
Perseids	Jul 25–Aug 17	Aug 12	50
Kappa Cygnids	Aug 18–22	Aug 20	(5)
Orionids	Oct 18–26	Oct 21	20
Southern Taurids	Sept 15–Dec 15	Nov 19	5
Northern Taurids	Oct 15–Dec 1	Nov 19	(less than 5)
Leonids	Nov 14–20	Nov 17	(5)
Geminids	Dec 7–15	Dec 13	50
Ursids	Dec 17–24	Dec 22	15

meteor showers (Leonids).   These interesting meteors are the remnants of the famous showers seen in the past. They have been known since A.D. 902 and are associated with Tempel's Comet of 1866. They travel in retrograde orbits. The meteors which reach their peak on November 16 are swift with an unpredictable hourly rate—some 30,000 per hour were seen in 1833. Heavy showers also occurred in 1866, 1898, and 1901. In recent times rates have been low, but in 1961 there was a definite revival which increased over the succeeding years and culminated in the magnificent 1966 display when observers in the United States recorded thousands of meteors per hour. The extraordinary concentration of these meteors is indicated by noting observerers in Europe whose dark skies a mere 9 hours before the Leonids greatest intensity recorded rates of only 2 or 3 a minute.

meteor showers (Perseids).   The best known and most reliable of all the annual meteor showers. The meteors usually reach their maximum on the night of August 11/12 when up to 60 swift meteors an hour may be seen. Early Perseids are seen in mid-July and persist for about a month, with the radiant shifting northeastward from night to night—beginning in Andromeda and finishing up in Camelopardus. This is an effect of parallax caused by the Earth's motion through the swarm. These "August" meteors have been recorded for at least one thousand years.

meteor stream.   A group of meteoric bodies with nearly identical orbits. The term is sometimes used to designate the objects causing a meteor shower.

meteor train.   Luminous linear streak left along the trajectory of the meteor after the meteor has passed. It may be visible against a dark sky for periods ranging from a few seconds (ionized trains) to many minutes (smoky or dusty trains) and may become contorted by high-altitude winds.

meteorites.   Solid objects that strike the ground after entering the atmosphere from interplanetary space. There are three broad classes: stony meteorites (about 93 percent of all falls) are silicate rocky material: iron meteorites (about 6 percent) are nearly pure nickel–iron alloy: stony-irons (about 1 to 2 percent) are stony types intermixed with nickel–iron material. Geological and geochemical analysis indicates that they formed inside parent bodies resembling asteroids, typically some hundreds of km in diameter, about 4.6 billion years ago. They are believed to be fragments of these bodies broken apart by collisions among themselves during the formation of the solar system. Further evidence indicates that they are related to the actual asteroids observed today. Spectral studies show that their spectra resemble those of certain asteroids, especially the Apollos whose orbits cross Earth's orbit and could actually intersect the Earth. Orbits observed for two or three meteorites resemble Apollo-type orbits and reach into or across the asteroid belt, suggesting that some of these bodies may have been somehow deflected out of the belt, eventually to fall on earth. Most meteorite material has been modified from primitive chemical forms by geological processes such as heating and exposure to pressure; some have been entirely melted and resolidified. However, a subtype of the stones known as carbonaceous chondrites (about 6 percent of all falls) appears to be very primitive material containing carbon and water never strongly melted. It may represent surviving material from the matter that formed in the ancient solar system and aggregated into planets.

meteoroids.   Small bodies in interplanetary space that would produce meteors or meteorites if they struck Earth. The pre-meteor types may have very low densities (0.05 gm/cm^3) based on rocket sampling measures, and may consist of dust and ices. Radar observations indicate that they may disintegrate prior to aerodynamic heating when exposed to the relatively small acceleration in the high atmosphere.

meter. The fundamental unit of length in the metric system equal to 39.37 inches. It is now defined in terms of atomic processes as 1,650,763.76 wavelengths of radiation from a specified electron transition in a krypton-86 atom.

metonic cycle. A period of 19 years, after which the various phases of the Moon fall on approximately the same days of the year as in the previous cycle, also equals 235 lunations. It was discovered by the Greek astronomer Meton around 430 B.C.

microbarograph. An instrument of great sensitivity used to record the atmospheric pressure.

micrometeorites. Meteoroids too small to be observed by the conventional technique. They survive passage through the atmosphere, ultimately settling slowly on the Earth at the estimated rate of 10^7 kg per day, and so they are in a sense meteorites. The number of micrometeoroids increases rapidly as the particle size decreases. Although the masses of the micrometeoroid particles are extremely small, they travel in space with high geocentric velocities (around 12 to 72 km per second) and may be mostly comet debris.

micro-meter. 10^{-6} meter, abbreviated μm. A device in the eyepiece of a telescope for measuring the angular dimensions of a celestial object.

micron. The micron is a thousandth of a millimeter or 10^{-6} meter, and is designated by the symbol μ (Greek mu). It is a useful unit for expressing the wavelength of infrared waves. The term is being replaced by micro-meter or μm, according to the International System of Units (q.v.).

Microscopium (The Microscope). Below Capricornus lies a small faint constellation whose name, Microscopium, indicates that it is a relatively modern addition to the sky, first being drawn by Lacaille in 1752. It has an area of 210 sq deg immediately south of Capricornus, with no conspicuous feature for recognition. There are no objects of telescopic interest beyond a few double stars and the galaxy NGC 6925.

microwave. Radio radiation having wavelengths between 1 mm and 30 cm.

Mie scattering. Scattering (or dispersion) of light by diffraction from small particles such as dust particles or droplets in an atmosphere.

Milky Way. The appearance of our own galaxy as seen from our position inside its disk, which extends completely around the entire sky as a hazy band of light. Its width averages about 20°; its middle forms a great circle of its own in its heavens, cutting both equator and ecliptic. One way of specifying the position of the Milky Way is by the coordinates of its northern pole, right ascension 12^h49^m, declination $+27\ 1/2°$ (1950 coordinates). In addition to its great brightness in a very dark rural sky, a striking feature of this luminous belt is the local variations in dark and bright shades throughout its entire circuit. Aside from the long dark rifts that apparently divide the stream, it is mottled and crossed in many places by dark areas which correspond to dark dust clouds silhouetted against more distant stars. Because of its inclination, the appearance of the Milky Way is quite different at different hours of the night and from night to night throughout the year. At nightfall in the late summer in middle northern latitudes, the Milky Way arches overhead from the northeast to the southwest horizon. It extends through Perseus, Cassiopeia, and Cepheus as a single band of varying width. Beginning in the fine region of the Northern Cross overhead, it is apparently divided into two parallel streams by the Great Rift, which is conspicuous as far as Sagittarius and Scorpius. The western branch of the Milky Way is the broader and brighter one through Cygnus. Farther south, in Opiuchus, this branch fades and nearly vanishes behind the dense dust clouds, coming out again in Scorpius. The eastern branch grows brighter as it goes southward and gathers into the great star clouds of Scutum and Sagittarius, which lies in the direction of the galaxy's center at about 17^h42^m, $-28.9°$. The galactic nucleus itself is obscured by dust, but the bulged bright region around the center can easily be seen by the naked eye in the Sagittarius area. In the evening skies of the late winter in middle northern latitudes, the Milky Way again passes nearly overhead, now from northwest to southeast. The stream is thinner here and undivided. From Cassiopeia to Gemini it is narrowed by a series of nearby dust clouds which cause a pronounced obscuration north of Cassiopeia and angle down through Auriga to the southern side of the band in Taurus. The Milky Way becomes broader and weaker and less noticeably obscured as it passes east of Orion and Canis Major down toward Carina. The name goes back to the Latin *Via Lactea*, or Milky Way.

milli.   Prefix indicating one thousandth (10^{-3}).

millibar.   A widely used unit of atmospheric pressure. Equal to 1/1000 bar or 1000 dynes per square centimeter. One millibar = 0.0295299 inches of mercury. Typical Earth surface pressure is about 1000 millibars.

Mimas.   An inner satelite of Saturn orbiting just outside the rings at about 186,000 km. The diameter is estimated roughly at 300 km.

minute of arc.   An angle equal to 1/60 of a degree; therefore, an angular distance in the sky which is 1/21,600 of a circle; written 1′.

Mira (The Wonderful Star).   Mira appears and disappears with fair regularity. It has a period of about 331 days; at its brightest it may equal Polaris, but usually it is at about magnitude 8 to 10 (too faint to be seen by the naked eye). Mira, a long-period variable, is quite different from the regular-as-clockwork short period variables such as Delta Cephei. Its period is not precise, and the interval between successive maxima may not be exactly 331 days. Moreover, some maxima are brighter than others. When Mira brightens, it may be visible for a few months with maximum brightness ranging from magnitudes 2 to 4. (See Fig., p. 149.)

Mirach.   See Beta Andromedae.

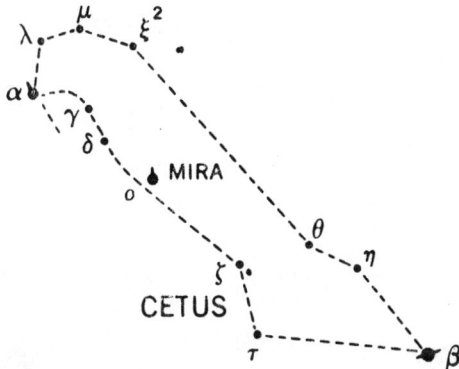

mirage.   A refraction phenomena in the atmosphere wherein an image of some object is made to appear displaced from its true position, or when light is refracted by inversion layers (q.v.) giving a glassy or watery appearance.

Miranda.   A satellite of Uranus orbiting at a mean distance of 128,000 km. Its estimated diameter is roughly 550 km.

Mirfak.   See Alpha Persei.

mixing ratio.   A measure of humidity. Equal to the mass of water contained along a given mass of dry air. Generally expressed in grams of water per gram or kilogram of dry air and designated as w or W.

Mizar.   The brighter component of the optical double Alcor and Mizar, middle star in the handle of the Big Dipper. While Alcor and Mizar are not truly co-orbiting, Mizar itself is famous as a system of 4 co-orbiting stars. The telescope reveals 2 discovered by Riccioli in 1650 and separated by 14″. In 1889 Mizar A was found to be a close binary by spectroscopic studies which revealed orbital motions by Doppler shifts. In 1908 Mizar B was also found to be a spectroscopic binary.

model atmosphere.   Theoretically calculated tabulation of temperature, surface gravity, pressure and chemical composition for a star, selected to reproduce the observed radiation of a given star.

molecule.   The smallest part of a compound that retains chemical identity with the compound. Defined as a combination of two or more atoms.

monocentric telescopic eyepiece.   An eyepiece made up of a triple lens system, useful for variable star work in that it gives very good definition, but having a relatively small field of view.

Monoceros.   Attributed to Bartschius in 1624 but probably defined much earlier, since Scalinger found it on a Persian celestial sphere of the previous century. It is an inconspicuous constellation occupying 482 sq deg between Orion and Canis Minor. The Milky Way, broad and diffuse here, passes its center. This region is very rich in newly formed stars and open clusters, at least 40 being known in this group. These include the remarkable NGC 2244 which is associated with a large nebulae NGC 2245 and 2261. Among the stars the brilliant white triplet β Mon is notable. A faint nebulous star, R Mon, is believed to be a newly formed star surrounded by dust possibly forming a planetary system.

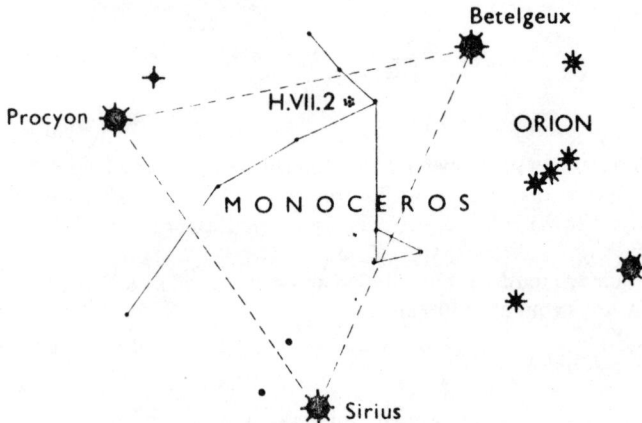

monochromatic.   Light of a single color, or a very narrow range of wavelengths, such as the light of a single spectral emission line. Special color filters can be used to photograph objects in just one color and thus give monochromatic images.

Monte Carlo calculation.   A type of computer calculation introducing sets of random or selected variables to represent plausible values of variables in highly complex problems.

Moon.   The natural satellite of the Earth. Early telescopic observers charted features such as the mare plains, rugged upland regions, and craters, but had little understanding of the origin of the features or the geology and chemistry of the surface. Dynamical studies gave data such as:

diameter	3476 km
mean distance	384,401 km
mass	7.35×10^{25} gm = earth mass
surface gravity	162.2 cm/sec^2 = 1/6 of earth's
mean density	3.3 gm/cm^3
escape velocity	2.38 km/sec

The first manned landings in 1969 and later landings and automated sample return missions allowed analysis of lunar rocks and measurements on its surface and gave us much better data on its physical nature. These measurements indicate that the Moon formed 4.6 billion years ago and was strongly heated during its first few hundred million years, causing melting and formation of igneous rocks in its outer layers. Much of the outer crust, still preserved in the cratered uplands, formed from a low-density type of rock known as anorthosite or anorthositic gabbro. Large impact basins and other depressions were flooded about 3.8 to 3.2 billion years ago by basaltic lavas generated beneath the surface, forming the dark mare plains that cover roughly 18 percent of the Moon. Few geological processes have occurred since as there is virtually no atmosphere or water to allow erosive processes, and as the interior has only modest seismic activity. Some moonquakes are generated by tidal forces exerted by the Earth. The meteorite impact today is modest (similar to Earth's), but was thousands of times greater in the first half-billion years of lunar history. The moon has few volatile elements (such as water and light gases) and lacks as much iron and iron-related minerals as the Earth. See lunar features.

moon eclipses.   See lunar eclipse.

Moon, Harvest.   See Harvest Moon.

moon phases.   The various shapes presented by the illuminated part of the Moon as it passes through its orbit around Earth: new, crescent, first quarter, gibbous, full, gibbous, last quarter, crescent, and new again. The changing shapes result from the varying angles at which the sunlit hemisphere is viewed from the Earth.

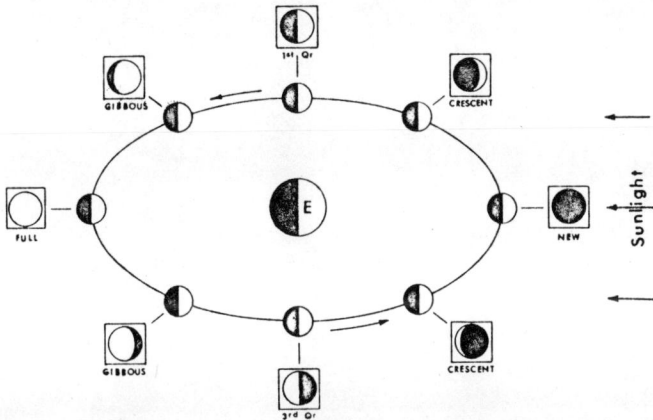

moonrise.   The crossing of the visible horizon by the upper limb of the ascending Moon.

moonset.   The crossing of the visible horizon by the upper limb of the descending Moon.

motions of the stars.   (1) The daily circling of stars around the sky, due to the real rotation of the Earth on its axis. (2) Movements of stars with respect to each other in three-dimensional space. In most cases these cause angular motions barely detectable from Earth, being only 10.3 seconds of arc/year in the maximum known case. See also proper motion.

M-region magnetic storm.   A solar magnetic storm that is independent of visible solar disk features.

Musca (The Fly).   Musca, the fly, was originally Apis, the bee, made by Bayer about 1603. In Lacaille's chart of 1763 it is designated by the modern name. It is a small constellation of area 138 sq deg lying immediately south of Crux, partly in the Milky Way and recognizable by a group of fairly prominent stars. Some fine telescopic objects occur in Musca. As well as several double stars there are two globular clusters, two interesting nebulae, and the remarkable diffuse nebula NGC 5189. Distant galaxies appear to be absent in this heavily veiled region.

N

Nacreous clouds. Rare clouds of the stratosphere exhibiting a remarkable play of colors. The colors extend, band after band, into the cloud center in distinction from the so-called irridescent clouds of the troposphere, in which the colors are arranged in fringes around the edges of the cloud. The cloud particles may be either ice crystals or subcooled water droplets. Nacreous clouds appear at heights of between 27 and 30 km above the surface of the Earth.

nadir. The point on celestial sphere opposite the zenith (i.e., directly beneath one's feet).

nano. A prefix meaning one billionth (10^{-9}).

nautical mile. The mean length along the meridian subtended by one minute of arc as seen from the center of the Earth, equaling 6,080 feet or 1.853 km.

nautical twilight. That period when the upper limb of the Sun is below the visible horizon and the center of the Sun is not more than 12° below the celestial horizon.

near-parabolic comets. Comets whose orbit shapes are close to parabolas. About three-fourths of newly discovered comets have orbits with eccentricities between 0.99 and 1.01. They are the near-parabolic comets. Because of the law of areas a body in a near-parabolic orbit sweeps through the solar system at unusually high speed; for a perihelion distance of 1 a.u. the speed is 42 km/sec. Because of such high speeds and because comets usually spend a good share of their visit in the daytime sky, it is sometimes difficult to obtain a number of positional observations, leading to some uncertainty in the orbits. Probably none have initial eccentricities actually exceeding 1.0, which would imply that they come from beyond the solar system. Near-parabolic comets are long-period comets; they differ from short-period comets, asteroids, and planets in having their orbits inclined at random and often high angles to the plane of the solar system. They probably come into the inner solar system from the "Oort cloud" of comets (q.v.), orbiting the sun far beyond Pluto.

nebula. Latin word for "fog." A nebula in space is an immense cloud of gas, mainly hydrogen, interspersed with particles of floating cosmic dust. As time passes, the cloud grows smaller and denser as a result of the mutual gravitational attraction of the gas molecules and dust grains.

nebulae. Gaseous clouds in interstellar space. At first this term was applied indiscriminately to all celestial objects of a hazy character which the eye could not resolve into stars, but many of these have since been found to be clusters of stars and galaxies. Faintly luminous objects are called "emission nebulae." Obscuring clouds of interstellar dust are often called "dark nebulae," the best-known example of which is the Great Nebula in Orion M42 which lies close to the Hunter's Belt. Some nebulae shine only by reflecting the light of stars contained in them, but others contain extremely hot stars whose radiation knocks electrons into high energy levels in the nebulae gas, exciting gas atoms and allowing them to radiate as the electrons return to their ground states, producing soft glows. Some nebulae are stellar birthplaces; inside them, fresh stars are being produced out of the nebular gas and dust. Other nebulae are debris of explosions marking disruptions of unstable stars.

nebulae, planetary.   See planetary nebula.

nebular hypothesis.   A theory on the formation of the Sun and planets (and by extension, other stars) that was first suggested in 1735 by the philosopher Immanuel Kant and then put in more precise astronomical terms in 1796 by the French astronomer and mathematician Pierre Simon Laplace. The original form of the theory presumed that the solar system was created from a vast cloud of gas. As the gas cooled and contracted, its speed of rotation increased to form a disk. The gases coalesced, the disk shrank, and a ring was abandoned by the contracting mass. Successively small rings were subsequently left behind, each forming a planet. The main mass, which contained more of the material, formed the Sun. The theory failed in its original form because it predicted a very rapidly rotating sun and did not explain the formation and spacing of the "rings" or planets in sufficient detail. However, modern theories of solar-system formation from a disk-shaped nebula are descendents of the original nebular hypothesis. See also planetesimal.

nebular variable.   A variable star found to be associated with either dark or bright nebulosity and including the RW Aurigae and T Tauri variables.

nebulium.   The hypothetical substance once thought to be the source of the emission lines in the spectra of gaseous nebulae. The lines were later found to be forbidden lines (q.v.) of known atoms.

negative eyepiece (telescope).   An eyepiece that forms a virtual image between the objective and the eyepiece.

nephoscope.   A meteorological instrument for measuring the direction and relative speed of clouds passing overhead.

Neptune.   Eighth planet from the Sun, orbiting at a mean distance of 30.1 A.U. Its extreme distance produces a very small angular size of about 2.1 seconds at opposition, too small to permit study of any detail. According to spectroscopic results, its atmosphere contains methane. Because of this and their similarity in size, the planet is believed to resemble Uranus in general. Neptune's mean density is about 1.6, somewhat exceeding Uranus's. Discovery of the Uranus rings in 1977 prompted speculation that Neptune might have a ring system in addition to its two known satellites, Triton and Nereid. Nereid is a small outer satellite, perhaps 540 km across. Triton is a large, retrograde inner satellite. Its diameter is highly uncertain, but it may be one of the largest satellites in the solar system at around 5000 km diameter.

Further data on Neptune include:

Diameter	49,500 km
Mass	1.0×10^{29} km
Rotation period	approx. 22 hours
Orbit inclination	1.8°
Orbit eccentricity	0.01
Escape velocity	N 25 km/sec

neutral point.   In atmospheric optics, one of several points in the sky for which the degree of polarization of diffuse sky radiation is zero.

neutrino astronomy.   Astronomical research based on observations of neutrinos. The neutrino is a strange fundamental particle of nature, electrically neutral with zero rest mass, traveling with the velocity of light and carrying energy. Neutrinos are believed to be produced in some reactions in stars (including the Sun). Neutrinos and antineutrinos are extremely difficult to detect because of their very great reluctance to interact with any form of matter. It has been calculated that on the average an antineutrino would travel a distance of 100 light-years through water before being captured by a proton. Tests to detect solar neutrinos in the 1970's led to a detection rate much lower than predicted, indicating that theories of the solar interior reactions or conditions may have to be somewhat modified. Another potential application of neutrino astronomy suggests that the production of neutrinos might play an important role in the development of supernovae and other unstable conditions in stars.

neutron.   A neutral particle with approximately the same mass as a proton; a component of atomic nuclei.

neutron star.   A collapsed star of higher density than a white dwarf, consisting primarily of neutrons, and formed after nuclear fuels are exhausted. Diameters would be about 10 to 15 km and densities about 10^{13} to 10^{15} grams/cm^3. Pulsars are believed to be rapidly rotating neutron stars.

neutrosphere.   Term occasionally used for the electrically neutral region of the atmosphere extending from the Earth's surface to approximately 75 km.

new moon.   The Moon at conjunction, when little or none of it is visible to an observer on the Earth because the illuminated side is away from the Earth. The Moon lies between Earth and Sun.

newton.   A metric unit of force equaling 10^5 dynes. The force needed to impart to 1 kg an acceleration of 1 m/s^2.

Newton, Isaac (1642–1727).   The father of mechanical physics, Newton was able to combine the various ideas of Copernicus, Kepler, and Galileo into a cohesive mathematical theory of motions and forces and explain the behavior of the planets and the forces that influence them. At age 24 he made several valuable contributions to science by developing the binomial theorem and differential calculus, and enunciating the theory of colors. Another of Newton's early contributions was the construction of a telescope which made use of mirrors instead of lenses to gather and focus light. Newton's greatest triumph is generally recognized as his Universal Law of Gravitation, which describes the gravitational forces between objects of any specified masses and separation distances.

Newton's Laws.   (1) A body remains at rest or continues to move with constant velocity in a straight line unless it is acted on by a net outside force. This fundamental property of all matter is called inertia. (2) The acceleration (change of speed and direction) of a body is directly proportional to the external force producing the change, inversely proportional to the mass of the body, and in the direction of the external force. (3) As two bodies interact, for every force acting on one body, an equal force in the opposite direction acts on the other body.

Newton's law of gravitation.   Every particle of matter in the universe, m_1, attracts every other particle (such as m_2) with a force, F, acting along the line joining the two particles, proportional to the product of the masses m_1m_2 of the particles and inversely proportional to the square of the distance r between the particles, or $F = Gm_1m_2/r^2$, where G = gravitational constant.

Newton's Principia.   Newton's *Philosophiae Naturalis Principia Mathematica* or the *Mathematical Principles of Natural Philosophy*, often called the *Principia* is regarded as one of the greatest scientific classics. In it Newton summarized the basic principles of celestial mechanics, developed the necessary mathematical procedures, simplified the methods, and demonstrated the nature of motions in the universe.

Newton's rings.   Circular interference fringes, or colored bands, seen when two nearly flat surfaces are in contact, so-called because Sir Isaac Newton first observed them. They appear because the light passes through the thin layer of air between the glass plates and reflects back and forth between the two glass surfaces. The fringes appear when the light is bent just enough to make the waves interfere with each other.

Newtonian reflector.   The world's first reflecting telescope was constructed in 1668 by Isaac Newton. The principle of the Newtonian reflector is that light from the star falls on the curved front surface of the main mirror which reflects it back up the tube. Before it forms an image, however, it is again reflected, through a right angle, by a second mirror (this time plane) known as the flat. The light then passes through a hole cut in the side of the tube and comes to a focus where the eyepiece is positioned. This arrangement keeps the eyepiece assembly from obstructing the main mirror. Thus, instead of looking "up" the tube, as with a refractor, the observer peers into the side. (See Fig., p. 157.)

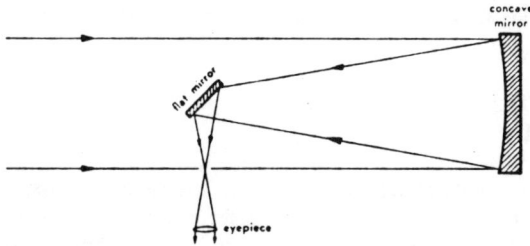

NFK—Neue Fundamental Katalog. One of the earliest modern "fundamental" catalogues of star positions, epoch 1870 and 1900.

NGC (followed by a number). Assignment given to an object listed by number in Dreyer's New General Catalogue of Nebulae and Clusters of Stars.

nightglow. A faint glow of the night sky. It is an atmospheric phenomenon caused by emission of light from excited atomic oxygen and other atoms at high altitudes. Nightglow (or airglow) probably occurs in some form at all times, both day and night, but is most readily observed and studied at night when the sky is dark. Although the nightglow spectrum has some features in common with those of auroras, it does not depend greatly on latitude or on exceptional solar activity. The Sun's normal radiations are ultimately responsible for the observed luminosity by exciting atoms on the illuminated side of Earth. Nightglow is faintest at the zenith and increases in intensity toward the horizon with its maximum about 10 degrees above the horizon.

nimbostratus. A cloud type described as "a low, amorphous layer of a dark gray color and nearly uniform; feebly illuminated, apparently from the inside. This cloud is associated with steady rain or snow." It is a low layer, uniform in composition, and of little outline. Sometimes in the lower side, broken patches (called "scuds") drift along at greater speeds. These are often seen when the rain is clearing. Quite frequently altostratus and nimbostratus merge into a single cloud layer. Nimbostratus is associated with general storm conditions extending over a wide area. These thick and heavy clouds are easily seen and recognized from the top. The upper surface shows vertical development and it is sometimes found to be associated with cumulonimbus on top.

noctilucent clouds. Clouds of unknown composition (possibly meteoritic dust) occurring at great heights (75 to 90 km) and visible at twilight. They resemble thin cirrus, but are usually of a bluish or silverish color (although sometimes orange to red) standing out against a dark night sky. Since they are illuminated by a sun 5° to 13° below the horizon these clouds are rare; they have been observed during summer months in both hemispheres in some parts of the latitude belts between latitudes 50° to 75° N and 40° to 60° S.

nodal (or nodical) month.   The period of the Moon around the Earth from one node back to the same node, 27.21 days.

nodal year.   The period of the Earth around the Sun with respect to the line of the nodes of the Moon's orbit.

nodes.   (1) The points at which the orbit of any satellite crosses the plane of the primary's equator or other fundamental reference plane such as the ecliptic. The movement of these crossing points caused by perturbations is referred to as regression of the nodes. (2) Points with zero amplitude along a standing wave. See also next entry.

nodes of the month (lunar nodes).   The nodes of the Moon's path are the two opposite points where it intersects the ecliptic. The ascending node is the point where the Moon's center crosses the ecliptic from south to north; the descending node is the point where it crosses from north to south. Regression of the nodes is their westward displacement along the ecliptic, just as the equinoxes slide westward in their precessional motion, but at a much faster rate; a complete revolution of the nodes of the Moon's orbit is accomplished in 18.6 years.

noise.   Random responses in observational equipment as opposed to signal; the measurable responses from the object being observed. Sources of noise include sky light, electronic fluctuations, etc.

non-gray atmosphere.   An atmosphere in which the absorption coefficient varies with wavelength. See also gray atmosphere.

nonthermal radiation.   A class of radio waves usually involving high-speed electrons moving in strong magnetic fields. The nonthermal radio sources are of three or more kinds: (1) remnants of supernovas and filamentary nebulae (2) normal external galaxies (3) radio galaxies.

noon.   The instant when the Sun's center crosses the meridian.

Norma (The Square).   A small constellation of 165 sq deg lying south of Scorpius between Ara and Lupus, introduced by Lacaille in 1752. Its immersion in the Milky Way prevents any distant galaxies from appearing in it. There are a few double stars, and although 14 open clusters are known, many are not effective in the telescope. There is one moderately bright globular cluster and 4 small planetary nebulae.

north celestial pole (NCP).   The point towards which the Earth's northern axis of rotation points. Northern stars therefore appear to rotate about this point. The 2nd magnitude star Polaris lies about one degree from the pole. The NCP's elevation angle above the horizon equals the observer's latitude.

north polar sequence.   A list of stars near the north celestial pole arranged in order of photographic magnitudes and used as reference stars in stellar photometry.

north pole.   In astronomy, that end of the axis of rotation of a celestial body which lies in the hemisphere centered on the north celestial pole.

Northern Cross.   The prominent cross composed of the 5 bright stars of Cygnus. Since it is part of the constellation Cygnus, this asterism is not listed as an official constellation.

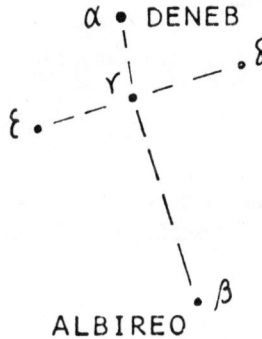

nova (plural, novae).   Originally, any temporarily appearing new star, including objects that would now be classed as supernovae. In the 20th century, the term has come to be restricted to a temporarily brightening star which follows a certain pattern of brightness change as a function of time. They brighten by as much as 14 magnitudes (often in a day or less) to absolute magnitudes of about -6 to -9. A long, slow fading to the original state then follows, often requiring years. Spectral studies show that the brightenings involve explosions in which the star throws off roughly a hundredth of a percent of its mass (contrasting with the much larger explosions of supernovae). Studies in the 1960's and 70's showed that novae are members of binary star systems in which the companion star is a red giant. It is believed that novae explosions occur when outer atmospheric gas escapes from the giant and falls onto the surface of its nearby companion—a dense, collapsed star resembling a white dwarf. The infusion of fresh hydrogen on the surface of the dwarf causes new nuclear reactions resulting in explosive activity and expulsion of gas from the system causing the nova.

novae, rapid.   These stars show an extremely steep rise to maximum which is followed by a fairly sharp decline that may then become shallower. In some cases the decline is marked by quite pronounced fluctuations, for example Nova V603 Aquilae (1918) and Nova GK Persei (1901). In others the decline is comparatively smooth, as was the case for Nova XX Tauri (1927) and Nova V476 Cygni (1920). The fast novae are generally regarded as those in which there is a decline of 3 magnitudes or more within 100 days of maximum.

novae, recurrent.   Novae-like stars with repeated sporadic explosive activity. Within the galaxy as a whole, some twenty or thirty novae occur each year. If, as believed, only certain types of stars pass through the novae phase, such stars must explode on more than one occasion consistent with recurrent nova activity.

novae, slow.   Unlike the rapid novae which have amplitudes of between 11 and 13 magnitudes, slow novae are somewhat smaller, usually between 9 and 11 magnitudes. A few of the slow novae, like Nova HR Delphini (1967), have even smaller amplitudes. Applying the same criterion as for the fast novae we can classify the slow novae as those in which the initial decline of 3 magnitudes takes longer than 100 days. The stay at maximum often lasts for several months and frequently there is a subsidiary rise following the decline to minimum.

Nubeluca Minor.   The Small Magellanic Cloud. Rarely used.

nucleon.   A neutron or proton, the major particles of an atomic nucleus.

nucleosynthesis.   The formation of nuclei of atoms (usually implying atoms heavier than hydrogen or helium) generally by reactions inside stars in which light nuclear particles are fused under high temperature and pressure into heavy nuclei.

nucleus.   (1) The relatively large central particle in an atom. The two basic units—protons and neutrons—that go into the makeup of nuclei can be called by the general name nucleons. Protons and neutrons are very nearly identical in their masses. They differ primarily in that the proton carries a unit of positive charge, whereas the neutron is neutral. Protons and neutrons are approximately 1840 times more massive than electrons. (2) The central bright condensation in many galaxies, especially spirals. (3) The central bright condensation in a comet.

nutation.   A slight nodding of the Earth's axis due to the precession of the axis and the lunar attraction on the Earth's tidal bulge superimposed on the larger precessional motion. The path of the north (or south) celestial pole on the celestial sphere due to precession is not exactly a circle because a small periodic disturbance is associated with the regression of the nodes of the lunar orbit. This cyclically changes the relative lunar and solar forces on the Earth's equatorial bulge, thus altering the precessional forces on Earth. This produces a regular variation in the precessional cycle. The effect is called nutation (Latin: *nutare*, "to nod") because Earth's rotational axis appears to nod back and forth as it precesses.

O

OAO. Abbreviation for orbiting astronomical observatory.

OB association. A grouping of O and B stars, somewhat looser in structure than an open cluster. Since O stars last only a few million years, the associations are young and associated with star formation.

oases. Small dark spots on Mars once said to mark junctions of "canals." Most are probably dark concentrations in the variable, windblown Martian dust deposits.

Oberon. A satellite of Uranus estimated to be about 2200 km in diameter.

objective. The principle lens or mirror in a telescope.

objective grating. See objective prism.

objective prism. A large diameter, small angle prism placed in front of the objective to give low dispersion spectra of all the stars in the field. A grating (objective grating) may be used for the same purpose.

oblate. Flattened at the poles, giving a nonspherical or noncircular shape.

oblateness. The departure of a planet from spherical form because of the centrifugal force of rotation. If equatorial diameter is a and polar diameter is b oblateness is given by $(a-b)/a$.

oblate spheroid. The nonspherical shape assumed by a spinning sphere due to centrifugal forces.

oblique projection. A map projection with an axis inclined at an oblique angle to the plane of the equator.

obliquity. The angle between a planet's equatorial plane and its orbital plane. i.e., the "tilt" of its polar axis.

Obliquity of the Ecliptic. The Earth's obliquity, $23.°45$.

obscuration. Absorption of light from distant stars by interstellar dust.

observed. Pertaining to a value which has been measured in contrast to one which is computed.

Occam's razor. See Ockham's razor.

occultation.   The disappearance of a celestial body through the passage of a nearer one in front of it; for example, occultation of a star by a planet or a planet by the moon. Thus a solar eclipse is strictly an occultation of the Sun by the Moon.

occluded fronts.   A weather system that occurs when a cold front overtakes a warm front, forcing the warm air aloft.

Ockham's razor.   The principle that the best hypothesis is the one with the fewest ad hoc assumptions, developed by William Ockham in the 1300's.

Oceanus Procellarium (Ocean of Storms).   The largest mare on the Moon, with an area almost twice that of the Mediterranean Sea; it lies near the northeastern limb and is a plain formed by lava flows.

Octans (The Octant).   A constellation established by Lacaille in 1752 to commemorate the invention of the octant by John Hadley in 1730. It occupies the rather barren region round the south celestial pole. There is little to distinguish this comparatively small area of 291 sq deg.

ocular.   An eyepiece.

Olbers' paradox.   The paradox, stated by German astronomer Heinrich Olbers in 1826 (and earlier by others) that if the sky is uniformly populated with stars to indefinite distances, the night sky should be as bright as the surface of an average star, due to the geometry of overlapping star disks as seen from the Earth or other points. The solution to the paradox is now believed to involve the recession of distant galaxies and the finite age of the big bang (of the order of 16 billion years). Because of the recession and red shift, we receive virtually no light from beyond a certain distance, and because of the finite age, the radiation in the universe may not have come into equilibrium with the matter.

Omega Centauri.   A prominent globular cluster in the constellation Centaurus appearing to the naked eye as a fuzzy star of the 4th magnitude. It is known as one of the most beautiful telescopic globulars in the sky. The stars composing it range from about 13th to 15th magnitudes and seem innumerable. It is among the nearest of the great globular clusters, about 16,000 light-years distant and about 65 light-years across.

Omicron Ceti (Mira).   See Mira.

Oort cloud.   A swarm of comet nuclei believed to populate the region from the outermost planets (30 to 40 A.U.) to a distance as great as 100,000 A.U. from the Sun.

Oort's constants.   Constants that characterize the rotation of the galaxy as seen from the Sun. Oort's constants characterize the apparent relative velocities of other stars as a function of distance from the Sun's orbit.

opacity.   A measure of a material's efficiency at stopping the passage of light rays.

open cluster (once called galactic cluster).   A loose grouping of stars, typically numbering 100 to 1,000, lying near the plane of the galaxy. Easily visible examples include the Pleiades and the Hyades, as well as many stars in the region of the Big Dipper. Most open clusters contain relatively young, stars; typical cluster ages are tens of millions of years. Stars are believed to form in open clusters following gravitational contraction and break up of large interstellar clouds containing hundreds of solar masses of material.

Ophiuchus (The Serpent Bearer).   An ancient constellation to which Ptolemy assigned 24 stars. It is a large group of area 948 sq deg stretching from Hercules to Scorpius on both sides of the equator; Serpens appears on each side of it. The brighter stars are widely scattered without pattern and this may be the reason why the compelling figure of Scorpius with a very short inclusion of the ecliptic is in the zodical sequence while Ophiuchus with an ecliptic passage nearly three times as long is not. The southern part of Ophiuchus is in the Milky Way, the center of which lies in this direction near the junction of Scorpius and Sagitarius. The region is heavily obscured with different nebulous matter, both luminous and dark. There are 6 open star clusters but none make an effective telescopic object. Many planetary nebulae are known.

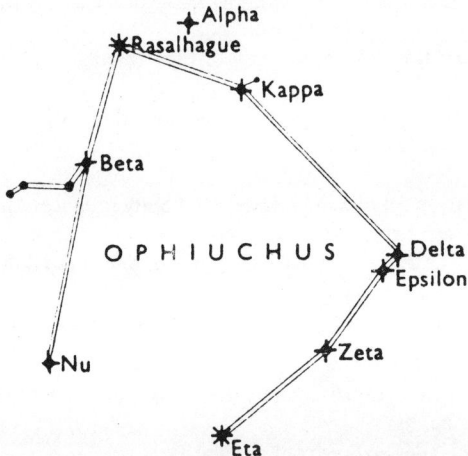

opposition.   The position of the Moon or a planet when it is in a direction diametrically opposite to that of the Sun. At opposition the Earth lies between the body and the Sun.

optical air mass.   See air mass.

optical axis.   The perpendicular to a tangent drawn at the center of a lens or mirror, containing the focus.

optical depth.   A dimensionless measure of the absorption of radiation in a given layer of material. An optical depth less than 0.1 indicates relatively clear passage of light; greater than 1.0 indicates a relatively opaque material.

optical double star.   Two stars which as seen from the Earth lie in almost the same direction and therefore appear to be close together and associated; a chance alignment of two stars not physically connected.

optical line of sight.   The path of visible light through the atmosphere, generally curved slightly by refraction.

optical sensors.   Electronic devices which determine the direction of the Sun or other luminous body.

optical window.   Portions of the spectrum in which a medium (such as the atmosphere) is relatively transparent. Absorption of radiation by Earth's atmosphere takes place over wide ranges of wavelength, leaving only two "windows" through which the Sun and stars can be observed: the visual and radio windows. From roughly 0.3 to 1 micron in the near-infrared wavelength is the visual window where the atmosphere is nearly transparent. The term optical window may be used to refer specifically to this visual window. By observing from above the atmosphere, space astronomers avoid the limitations of windows and can see nearly the whole spectrum.

optical path.   The path followed by a ray of light through an optical system.

optical train.   The series of lenses, mirrors, and any other optical devices constituting a complete specified optical system.

orbit.   The imaginary path of a moving celestial body in space. The term often refers specifically to a closed orbit such as an ellipse or circle, but can also apply to a nonclosed path, such as the parabola followed by an object moving at escape velocity.

orbital elements.   The seven quantities that must be established in order to specify the path of an orbiting body and its position in the path. These are:

(1) semimajor axis (a)—approximately mean distance from the Sun in the case of solar system planets;
(2) eccentricity (e)—departure from circularity;
(3) inclination (i)—or orbit plane to some other reference plane such as the ecliptic;
(4) longitude of the ascending node—where body crosses reference plane moving south to north;
(5) argument of periapse—where body is closest to the Sun or other central body;
(6) epoch (T)—moment of a passage through periapse point;
(7) period (P)—not required in case of planets, since it is determined by the Sun's mass; but required to specify orbit in an arbitrarily chosen system such as a binary star, where the orbiting masses are not necessarily known.

Orion (The Hunter).   A major and familiar constellation centered on a region of gas and recently formed bright stars. Ptolemy assigned 38 stars to it, including the nebulous one. The area is 594 sq deg. Many of the bright Orion stars are very hot, young, and of early spectral type. Orion is especially rich in beautiful double and multiple stars, and nebulous haze both bright and dark is found in many regions. There is no globular cluster, but 10 young open clusters are known, some of which are brilliant scattered groups. Orion is conspicuous in that it has two 1st-magnitude stars—the blue-white Rigel (Orionis) and the red star Betelgeus. This pair is about 20° apart, separated by a row of three 2nd magnitude stars 1 1/2° apart known as the Belt of Orion (less commonly, Jacob's Rod or the Yardstick). As it is situated on the celestial equator, the Belt rises at the east point and sets at the west. Running southward from the Belt are three fainter stars—the Sword of Orion—the central one of which appears a trifle hazy. Actually, this "fuzzy" spot in the sky is a vast volume of luminous gas and dust, the Great Nebula of Orion, which is the core of a star-forming region about 1500 light-years away. The view toward Orion is a view nearly along our spiral arm of the galaxy, or a dense spur off that arm.

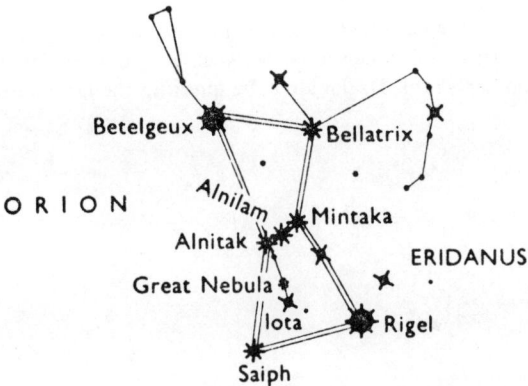

Orion arm.   The spiral arm of the Milky Way containing the solar system.

Orion spur.   A branch of our spiral arm that juts off from the Orion arm in the direction of the constellation Orion. A star-forming region in that direction.

orthoscopic eyepiece.   One of the highest quality eyepieces with respect to flat field and wide range of magnification. It contains two achromatic elements and has a flat field which is 45° to 50° side. Internal reflections are broad and offer little interference with the image produced. Field curvature and astigmatism are at a minimum and eye relief is excellent. It is especially useful for high powers.

outer atmosphere.   Very generally, the atmosphere at a great distance from the Earth's (or a planet's) surface, involving its thinnest outer parts.

outer planets. Jupiter, Saturn, Uranus, Neptune and Pluto.

overcast. A cloud layer which covers over nine-tenths of the sky.

overcoating. Transparent coating, usually of quartz or thin siliceous material, to protect the aluminum coating of a mirror.

ozone. A gas whose molecules consist of three atoms of oxygen (O_3). It is a minor constituent of the Earth's atmosphere, but of great significance, since it stops biologically-damaging ultraviolet light of the Sun. The first stage of formation of ozone (O_3) is the photochemical dissociation of oxygen molecules into atoms. An O atom may then combine with an O_2 molecule to form O_3.

ozone layer. Region in the Earth's upper atmosphere in which much of the ozone is concentrated. It lies at elevations of about 15 to 30 km.

ozonosphere. The ozone layer.

Ozma. A project in the early 1960's to listen for radio signals at 21 cm wavelength from intelligent aliens near other stars. No such signals were detected. Named by American astronomer Frank Drake after the queen in the Oz children's stories.

P

pair production. Spontaneous conversion of a gamma ray into an electron and a positron.

Pallas. The second asteroid to be discovered, and one of the largest, with a diameter of 610 km. It has an orbital period of 4.61 years, an exceptionally high orbital inclination of 34° 44′, and at its brightest a magnitude of 8.

Palomar Sky Survey. More correctly the *National Geographic Palomar Sky Survey*, a monumental photographic atlas of the heavens in two different colors covering the celestial sphere from the north pole to about −30° declination using the 48 inch (122 cm) Schmidt.

panspermia hypothesis. The hypothesis that life spreads by natural or artificial means from one planet to another, instead of evolving independently in each system.

parabola. A conic section formed by the intersection of a plane with a cone, the plane being parallel to the sloping side of the cone. It is the curve followed by an object moving at escape velocity.

parabolic velocity. escape velocity.

parallactic ellipse. The apparent ellipse followed by a nearby star relative to a background star due to the parallax effect of the earth's orbital motion.

parallactic inequality. A secondary effort in the solar perturbation in the Moon's longitude due to the ellipticity of the Earth's orbit.

parallactic motion of a star. The proper motion of the star due to the Sun's motion in space. It varies inversely as the distance of the star from the Sun.

parallax. An apparent shift in an object's direction due to a shift in the observer's position.

parallax of a star. The angle subtended at the star by the radius of the Earth's orbit or, equivalently, the semimajor axis of the star's parallactic ellipse. The smaller the parallax, the greater the distance. When expressed in seconds of arc, the parallax is numerically equal to the reciprocal of the distance expressed in parsecs. Thus a parallax of 0.1 second of arc (0.1″) corresponds to a distance of 10 parsecs. See also triangulation.

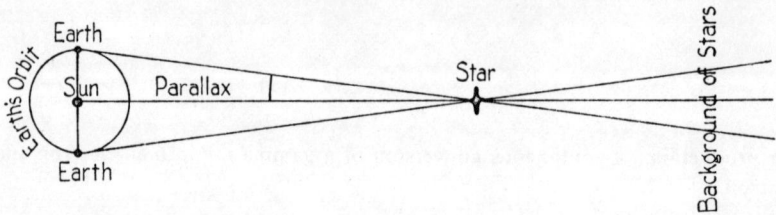

parallel.   (1) A circle on the surface of a planetary body or star, parallel to the plane of the equator and connecting all points of equal latitude. (2) A circle parallel to the primary great circle of a sphere or spheroid; also a closed curve approximating such a circle.

parallel of altitude.   A circle on the celestial sphere parallel to the celestial equator.

parallel of declination.   A circle on the celestial sphere, connecting points of constant declination.

parallel of latitude.   A circle connecting points of equal latitude on a planet or star.

parhelion.   A mock sun or bright spot at the same altitude above the horizon as the real Sun but to one side, caused by refraction in high-altitude cloud particles.

parsec.   Defined as the distance at which the stellar parallax is 1 second of arc. Astronomical distances beyond the solar system are so large that even the astronomical unit (the mean distance between Earth and Sun) is too small for convenience; hence the units employed are the light-years and the parsec. The parsec is 3.26 light-years, i.e., 3.09×10^{13} km, or 2.06×10^{5} A.U.

partial eclipse.   An eclipse in which the eclipsing body does not completely shadow or cover the body being eclipsed.

Paschen series.   A series of spectral lines arising from electron transitions beginning or ending at the 3rd energy level of hydrogen.

Pavo (The Peacock).   One of the constellations introduced by Bayer about 1603. It covers an area of 378 sq deg between Telescopium and the circumpolar Octans. The only conspicuous star is the second magnitude α Pav in the northern edge. The center culminates at midnight about July 13. The chief ornament of this constellation is the beautiful globular cluster NGC 6752. There are no open clusters, but two planetary nebulae are known and some 30 galaxies have been examined.

peculiar motion.   The velocity of a star with respect to the local standard of rest, so called because it is the velocity peculiar to the star itself, after solar motion has been subtracted.

peculiar spectrum. A spectrum that does not fit conveniently into the standard spectral classes.

peculiar velocity. Peculiar motion.

Pegasus (The Winged Horse). Pegasus is one of the ancient northern constellations to which Ptolemy assigned 20 stars. It is a very large group of area 1,121 sq deg extending from near the equator to 35°N and is recognized by the great square of moderately bright stars. It is conspicuous in the November skies, and can easily be traced because of the large square just a little south of overhead—one of the stellar landmarks. It is in the zenith at 10:30 P.M. in the middle of October, 9:30 on the first of November, and an hour earlier in the middle of the month. The northeast star of the square is Alpheratz, which marks the head of Andromeda, the remainder of the constellation extending toward the northeast. The stars composing the square are all bright stars of the 2nd magnitude, and are approximately 15° apart. Within the square at least 30 naked-eye stars are visible, and in a light, clear atmosphere over 100 have been counted.

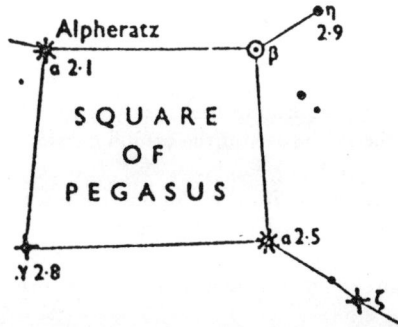

penumbra. The partly illuminated outer region of the shadow cast by a solid body from a light source of appreciable diameter. The region of semi-shadow over which the illumination gradually increases from total darkness to full illumination. From a point within the penumbra, the light source is partially but not totally occulted by another body.

penumbral eclipse. An eclipse in which the eclipsed object passes through a penumbral shadow. For example, when the Moon lies completely within the umbra, the eclipse is total; as it moves out of (or into) the penumbra, so that part is in the umbra and part in the penumbra, the lunar eclipse is partial. If the Moon lies further from the center of the Earth's shadow and does not enter the umbra, but only the penumbra, the phenomenon is referred to as a penumbral eclipse or a lunar appulse. (See Fig., p. 170.)

perfect cosmological principle. An hypothesis that the universe, on the large scale, looks about the same in all regions at all times. This was assumed in the steady-state theory of the universe, but seems disproved by evidence for the big bang theory.

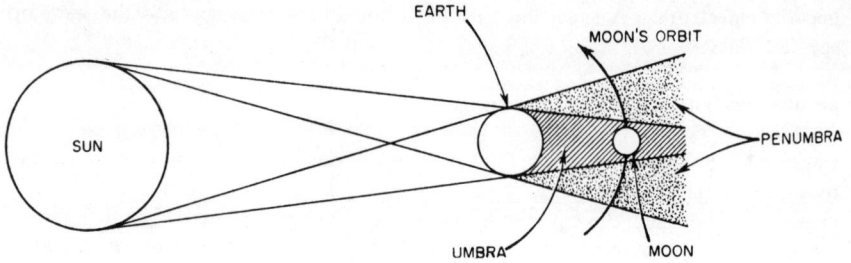

perfect gas.   See ideal gas.

peri.   A prefix meaning nearest.

periapsis.   The orbital point nearest the focus of attraction. Sometimes written periapse.

periastron.   The nearest point on the orbit of a stellar satellite to the star, or in a binary star the points in their orbits at which the two components are closest to each other.

perifocus.   The point on an orbit nearest the dynamical center (focus). The pericenter is at one end of the major axis of the orbital ellipse.

perigee.   The point on the orbit of the Moon or artificial satellite which is nearest to the Earth.

perihelion.   The point on the orbit of a planet or other orbiting body which is nearest to the Sun.

period.   (1) The time required for a body to make one complete orbit around its primary. The time interval of one complete circuit of an orbit or one complete rotation of a rotating body or the complete cycle of any periodic change. (2) The interval between successive maxima or minima of a variable star; the mean period in the case of a semi-regular variable being that averaged over a number of cycles.

period-luminosity relation.   An expression relating the periods of cepheid variable stars with their absolute magnitudes. Its discovery in 1912 by Henrietta Leavitt was important in allowing distances of remote variables in clusters and galaxies, out to about 3 Mpc, to be measured.

periodic comet.   A comet that returns to the vicinity of the Sun at more or less regular intervals; especially a short-period comet.

Perseids.   See meteor showers, Perseids.

Perseus (The Hero).   Perseus, the Greek hero is one of the ancient northern constellations. It lies in the Milky Way between Andromeda and Auriga from 31°N to 59° N, and may be recognized by a group of moderately bright stars low in the north preceding the brilliant Capella. The area is 615 sq deg. There are no very striking double stars in Perseus; no globular clusters occur and only two planetary nebulae are available for amateur instruments. There are few galaxies. However, Perseus contains the interesting eclipsing variable star, Algol. Normally of magnitude 2.3, it fades to magnitude 3.5 every $2^d 20^h 84^m$. The minimum brightness phase of the eclipse lasts only 20 minutes.

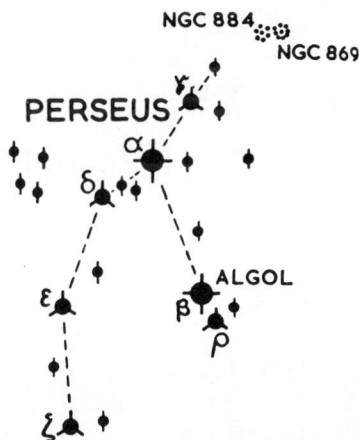

Perseus arm. The next outer arm of our galaxy beyond our arm; it lies about 12 to 13 kpc from the center.

personal equation. A systematic bias or error in observation associated with each observer's unique personality. The amount by which an observer's estimate of a quantity typically varies from the true value or the mean determined by many other observers.

perturbation. (1) A small gravitational force acting on an orbiting body from other bodies, causing disturbances in its normally Keplerian elliptical orbit. (2) Any small disturbance in a system, making it depart from an equilibrium state.

phase. The phase of the Moon or planet refers to the form of its visibly illuminated disk. The particular aspect of the Moon; the fractional part of a periodically occurring phenomena.

phase angle. (1) The angle sun–object–observer (i.e., the angle as seen from an object between the Sun and the observer). At full phase the phase angle of a planet is 0°; at quarter phase it is 90°; at new phase it is 180°. If the phase angle is divided by 180°, the result gives the fraction of the planetary hemisphere turned toward Earth that is not illuminated by the Sun. (2) A difference in phase (position the cycle) between two waves.

phase space.   A way of referring to multiple parameters to describe a particle or system. In addition to the normal parameters of x, y, and z to describe position, phase space may include other parameters such as velocity. It is thus an imaginary coordinate system with one axis for each parameter. A common example is 6-dimensional phase space with 3 position axes, x, y, z, and 3 velocity component axes, V_x, V_y, and V_z.

phases of the moon.   The various appearances of the Moon during different parts of the synodical month.

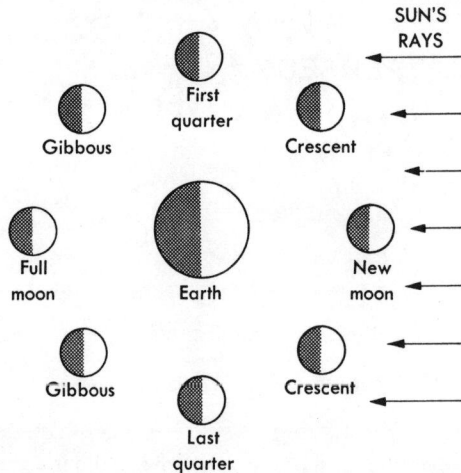

Phobos.   The inner of the two moons of Mars revolving in an orbit 5800 miles from the center of Mars (some 3700 miles above the surface) with diameter about 20 × 28 km (12 × 17 miles) and an orbital period of 7^h40^m. Because its orbital period is in the same direction as, but less than, that of Mars, it rises in the west and sets in the east as seen from Mars. It is heavily cratered and dark in color, of a material resembling carbonaceous chondrite meteorites. A system of grooves, possibly marking fractures, is associated with the largest crater, Stickner, which is about 10 km across.

Phoebe.   A satellite of Saturn orbiting at a mean distance of 12,960,000 km discovered in 1898 by E. C. Pickering. Its diameter is estimated to be 200 to 300 km.

Phoenix (The Phoenix).   One of Bayer's additions to the constellations about 1603. It lies south of Sculptor between Grus and Eridanus with little to distinguish its scattered stars. The area is 469 sq deg. The objects of interest in this constellation are restricted almost entirely to some double stars. (See Fig., p. 173.)

photoelectric cell.   A device for converting radiation, such as visible, ultraviolet, or infrared light, into an equivalent electric current; a type of transducer. Photocathode material which ejects electrons (called photoelectrons) when exposed to light. The

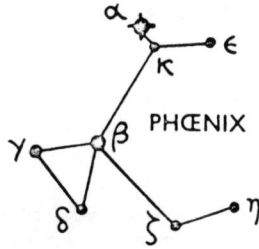

negative photoelectrons are attracted to a positive electrode (or anode) raised to some suitable positive voltage by a battery. The current induced in the circuit, amplified and measured, provides a determination of the amount of light falling on the photocell.

photoelectric effect.   Production of free electrons and voltage in some substance when struck by light.

photoelectric magnitude.   The magnitude of a star as determined with a photoelectric cell.

photoelectric photometer.   Sensitive instrument for measuring incident radiation using a photocell or a photo-multiplier tube. Used to measure brightnesses and brightness variations of faint stars, asteroids, etc.

photographic emulsion.   Modern photographic emulsion consists of a suspension of photosensitive microcrystals of a silver halide, usually silver bromide, in a thin gelatin layer on glass or acetate film. Astronomers mainly use glass plates, which give better geometric stability against stretching or distortion. Light chemically activates the grains so that the developer changes the silver halide to silver. Then the hypo (sodium thiosulfate), or fixing agent, dissolves away the underdeveloped silver halide, leaving black silver grains which form a negative image of the original bright image. Washing finally removes the hypo and its solute.

photographic infrared.   The portion of the invisible infrared region of the spectrum that can be photographed. It extends from about 7,600 Å to 12,000 Å. Most photographic emulsions are generally light-sensitive at shorter wavelengths than these.

photographic magnitude.   The brightness of a star as determined by measurements with photographic plates. These tend to favor blue light so that a blue star measures brighter than a red star of the same apparent visual brightness or total luminosity.

photographic telescope.   A telescope with an objective lens corrected best for blue wavelengths (to which photographic emulsions are sensitive) rather than for visual wavelengths.

photoionization. Ionization occurring when a photon of light knocks an electron out of an atom or molecule.

photometer. An instrument designed to measure the brightness of light falling on it. It may be visual (using the eye), photographic (using film) or photoelectric (using a photoelectric cell). A device for measuring the brightness of a celestial object.

photometric binary. An eclipsing binary discovered by the shape of photometric fluctuations in its light curve.

photometric function. The relation of brightness versus phase angle.

photometry, three-color. A system of making photometric measurements using three filters, one at a time, which allows measurement of light in three separate colors referred to as U (ultraviolet, 3600 Å), B (blue, 4200 Å), and V (visual, 5400 Å). Comparison of the brightness of a star in these three colors provides much information about the star and its spectrum. The system is often called the UBV system, or the Johnson-Morgan system, after its developers. In recent years, other filters with additional letter designations (R, red; etc.) have been added to extend the system to other wavelengths. See also photometry, four-color.

photometry, four-color. A variation on the UBV three-color system, often called the UVBy system. It uses filters passing ultraviolet (U), violet (V), blue (B), and visible (yellow, y) light.

photomultiplier. A photoelectric cell in which the sensitivity is greatly multiplied by allowing each original photoelectron to strike a target releasing even more photoelectrons which in turn release more, and so on.

photon. A single quantum of electromagnetic radiation having the properties of a particle. All electromagnetic radiations are emitted and travel from one place to another as photons. In a vacuum, all photons move with the velocity of light. Each photon carries a specified amount of energy proportional to the frequency (or wavelength) of the radiation. There is everywhere in the universe a certain photon density due to the cosmic background radiation and to diluted starlight that is the combined light emitted by stars in the universe. Since most photons are emitted by stellar gases whose temperatures range between 2000 and 20,000° K, their energies correspond mostly to wavelengths centered on the visible range.

photosphere. The visible surface of the Sun on which sunspots and other markings appear (Greek: *photo*, "light"). It is from this relatively thin part of the Sun's atmosphere (some 400 km in depth) that nearly all of the heat and light reaching Earth is radiated. The photosphere shows structure everywhere and is covered with a system of granules, giving it a mottled effect. A granule is about 1 second of arc across, only visible for a few minutes, and is the top of a convection current bringing up hotter material from below. The photosphere appears sharply defined because hydrogen ions in this layer make it much more opaque than the overlying clear gas. There is no sharp discontinuity in gas density at the photosphere, however.

photovisual magnitude.   A stellar magnitude determined from photographs using special orthochromatic or panchromatic emulsion and an isochromatic filter corresponding to the yellow or visual region of the spectrum. It is thus the magnitude perceived by the eye.

physical double star.   The co-orbiting stars, as distinguished from an optical double star (two stars in nearly the same line of sight that differ greatly in distance from the observer).

pico.   A prefix indicating a millionth of a millionth, or 10^{-12}.

Pictor (The Painter).   Pictor was originally called "The Painter's Easel" by Lacaille in 1752, but that name was simplified by Gould in 1877. It is a small constellation south of Columbia of area 247 sq deg and contains only scattered stars, making recognition difficult. Apart from some double stars, this constellation contains little of interest.

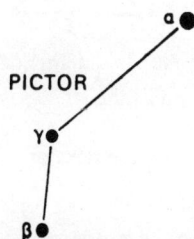

pion.   A neutral or charged particle with mass between that of an electron and that of a proton and with zero spin. Pions (also called pi-mesons) are highly unstable and believed to be involved in the forces that hold an atomic nucleus together.

Pisces (The Fishes).   Pisces is one of the zodiacal constellations and was assigned 34 stars by Ptolemy. It is a large group of 889 sq deg area in a rather barren region of the sky. It may be recognized by two irregular chains of faint stars, one roughly from north and the other from west, meeting in the pale yellow α Psc of magnitude 4.3 south of the triangle of Aries. The center of the constellation culminates at midnight about September 27. The only objects of interest in modest telescopes are a number of double stars and a few galaxies. (See Fig., p. 176.)

Piscis Austrinus (The Southern Fish).   Lies south of Aquarius and Capricorn in the region of the "The Sea" and is one of the ancient 48 constellations. In Greek mythology it was said to commemorate the transformation of Venus into the form of a fish when fleeing from the monster Typhon. It contains one bright star, the southern 1st-magnitude star Fomalhaut. This star is the mouth of the fish into which

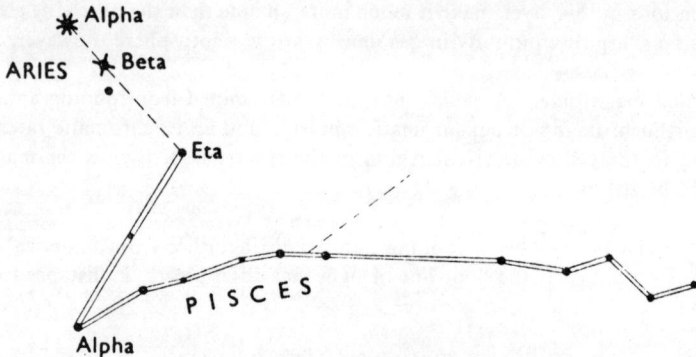

old star charts show the stream of water from Aquarius ending. The area of the constellation is 245 sq deg. There are many galaxies but few are prominent.

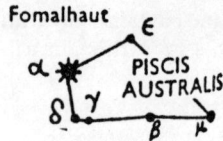

pits.   Tiny holes in the surface of a telescope mirror resulting from large particles of abrasive being present during the polishing.

pixel.   A picture element (i.e., one of the dots or resolution elements making up a digitized picture).

PKS.   A prefix followed by a set of numbers indicating an object in the Parkes Catalogue of Radio Sources.

plages.   (Latin: *plaga*, "region" or "zone") Also called "flocculi". The immediate bright "rims" in the neighborhood of sun spots. The temperature in a sunspot is only about 4000°K while the bright material in the plage area is chromospheric gas—usually a few hundred degrees hotter than its surroundings. The plage area usually appears some weeks or so before the first spots and may outlive the spots by several months.

plages, radio.   Solar areas apparently producing radio waves of 3 cm to 1 m wavelength. They are associated with visible plages and have lifetimes ranging from days to months, often lingering after plages disappear.

Planck's law.   A formula describing the amount of radiation at each wavelength coming per second from one square cm of a black body of any given temperature.

plane of the ecliptic.   The plane in which the Earth's center moves around the Sun. Often used as a reference plane in defining other planets' orbits.

planetarium.   A device projecting a replica of the heavens on an indoor dome which many people may view. The word planetarium refers either to the projection apparatus or to the structure that houses it.

planetary boundary layer.   The lowest layer of a planet's atmosphere, being the layer at which motions involve interactions with the surface. Above this layer lies the free atmosphere.

planetary circulation.   The system of large-scale disturbance in a planet's troposphere when viewed on a hemisphere or world-wide scale. The mean or time-averaged hemispheric circulation of a planetary atmosphere.

planetary configuration.   Apparent positions of the planets relative to each other and to other bodies of the solar system as seen from the Earth. Below is a diagram of a planetary configuration: (O, opposition; C, conjunction; IC, inferior conjunction; SC, superior conjunction; EE, eastern elongation; WE, western elongation; EQ, eastern quadrature; WQ, western quadrature.

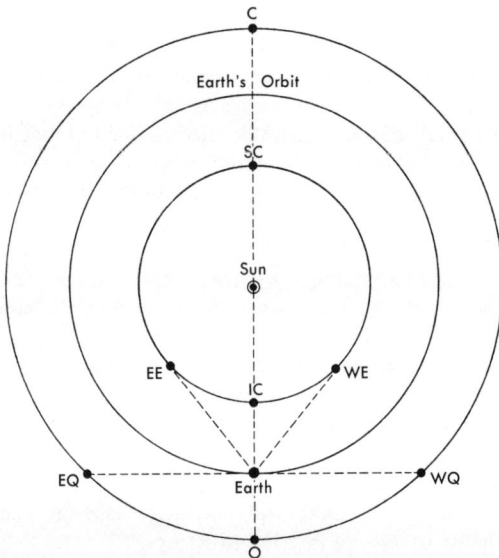

planetary interior.   The interior part of a planet, or a model thereof, usually described in terms of the temperature T, pressure P, density ρ, and composition at various depths from the surface to the center.

planetary nebulae.   Nebulae named by William Herschel because of the planet-like appearance which many of them show on the telescope (but having nothing to do with planets in reality). They appear usually as small round or somewhat elliptical bodies, often quite bright white or pale blue with fairly defined edges. Some of them look annular with a fainter central region which may be due simply to the effect of projection on a plane of a luminous but very tenuous gaseous spherical shell, like a bubble. Although visually they seem to shine with fairly even light, yet photographs with large instruments often reveal much irregular and sometimes filamentary structure which suggests turbulence. They range in apparent diameter from almost stellar points to objects 100″ or more across, the largest by far being the faint annulus NGC 7293 in Aquarius which is 15′ in diameter. They are produced when unstable stars (probably 0.6 to 4 solar masses) in their post-hydrogen-burning stages blow off their outer layers, creating expanding spherical shells. The remainder of the star collapses to form the white dwarf that generally marks the center of a planetary nebula.

planetary origin.   The process by which planets form—probably by aggregation of metallic, silicate, and icy grains that condense in the cool nebula around a newly-formed single star. By the action of gravity and collisions, these accumulate into spheres, the largest of which sweeps up most remaining small bodies. See solar system origin.

planetary precession.   That component of general precession caused by the effect of other planets on the equatorial protuberance of the Earth and producing an eastward motion of the equinoxes along the ecliptic. Planetary precession of the equinox is eastward at approximately 0.11 seconds of arc per year.

planetesimals.   The small bodies (generally microscopic to lunar in size) that were primordial precursors of planets and parent bodies of meteorites. They probably formed planets by aggregating, chiefly by gravitational attraction and collisional processes.

planetocentric.   (1) Of or pertaining to a planet's center of mass. (2) Of or pertaining to the planet as a center of a system; viewed from the planet's center.

planetographic.   Referring to positions on a planet measured in latitude from the planet's equator and in longitude from a reference meridian.

planetoids.   Old term for asteroids.

planetology.   The study of planets and satellites, especially in regard to the interpretation of their surface and interior structures.

planets.   (1) Solid or solid/liquid bodies, usually considered larger than 1000 km diameter, circling any star. (2) The nine large bodies circling the Sun in isolated orbits: Mercury, Venus, Earth, Mars, Jupiter, Saturn, Uranus, Neptune and Pluto (in order out from the Sun). The concept is not precise, since some other bodies are larger and since other bodies exist in sun-centered orbits.

planets, major.   The giant planets (q.v.) also called Jovian planets.

planets, orbits and properties.   See appendix.

planets, terrestrial.   Mercury, Venus, Earth and Mars, all of which have diameters, compositions, and densities relatively similar to those of Earth (in contrast to the giant planets).

planisphere.   A small-scale star map of the whole northern or southern sky with an oval or circular aperture that can be adjusted for any time and date to show which constellations are then above the horizon.

plano-convex lens.   A lens whose opposite sides are flat and convex.

plasma.   A gas consisting of ions and electrons free to move separately. Unlike a neutral gas, a plasma is affected by electric and magnetic fields, since it consists of charged particles.

Pleiades.   One of the best known groups of stars in the entire heavens, an open cluster visible to the naked eye in the shoulder of the Bull. These stars are celebrated in the legends of many ancient nations, and peoples far removed have attached to them ideas significantly similar. Six of these stars can easily be seen with unaided eye. The seventh star can be seen if one has very good eyesight, the cluster is known as the seven sisters. More than 100 stars can be seen with a small telescope. Astrophysical measurements indicate that the cluster is about 400 light-years away, roughly 13 light-years across, and about 50 million years old.

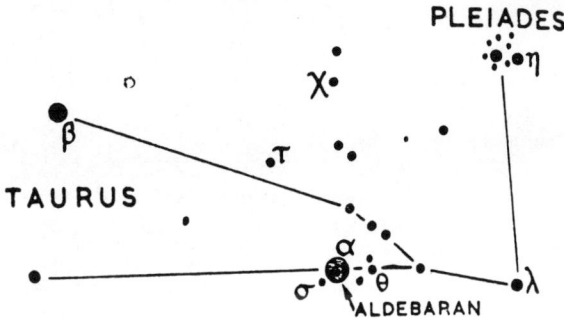

Pleone.   A fast-rotating bright star in the Pleiades known for its occasional production of shells of gas, probably thrown off its equator due to surface activity coupled with fast rotation.

Pluto.   The most distant known planet. The surface is mainly covered with methane frost, revealed by spectroscopic work in 1976. The semimajor axis of 39.4 astronomical units corresponds to a period of revolution of 248 years. The eccentricity of 0.25 and inclination of 17° are the highest e and i values for any planet. Because of the high eccentricity, Pluto crosses the orbit of Neptune, being the only planet with such a characteristic (but a collision in modern times is ruled out by the high inclination). Because of these properties, Pluto has been suspected of being not a true planet, but an escaped satellite of Neptune. In 1978, James Christy at the U.S. Naval Observatory discovered a satellite about 15,000–20,000 km from Pluto. Further analysis suggests a diameter around 2,100–3,000 km, a density around 0.7 to 1.5 g/cm^3 for Pluto, suggesting an icy composition. The satellite has roughly 0.4 times the diameter of Pluto.

point source.   A source of light of very small angular extent, usually less than a few minutes of arc.

polar axis.   The axis in an equatorial mounting parallel to the Earth's axis, thus pointing toward the celestial pole.

polar cap.   A bright or dark polar region on a planet or satellite. A famous example is the pair of bright polar ice fields on Mars. Polar caps are favored by the fact that temperatures at the poles are generally much lower than elsewhere, promoting condensation or freezing of atmospheric substances.

polar cap absorption (PCA).   A terrestrial ionospheric storm observed only in high latitudes, and accompanied by a blackout of long-range radio communication in these regions by the absorption of high-frequency radio waves. It is associated with solar flares, but may commence fairly rapidly, within 20 minutes to an hour or so after the flare, or more gradually about a day later.

polar distance.   Angular distance from a celestial, planetary, or stellar pole; the arc of an hour circle between a celestial pole, usually the elevated pole, and a point on the celestial sphere, measured from the celestial pole through 180°.

polar front.   The more or less permanent boundary between the cold polar easterly winds and the relatively warm southeasterly wind of the middle latitudes on earth. A similar phenomenon is seen on Mars.

polar hood.   A winter cloud layer extending outward from the poles, often seen on Mars and Earth.

polar orbit.   The orbit of satellite that passes over a planet's or satellite's pole.

Polaris.   The pole star, or nearest bright star to the north celestial pole. It is a relatively distant star, a supergiant about 780 light-years away. It is slightly variable, with a 4-day period. It has a visual companion. It is also a spectroscopic binary and may have two additional 12th magnitude companions. Polaris is not in exact line with the Earth's axis, but describes a small circle about the true Pole 2 1/2 degrees in diameter. Polaris will continue to approach the position of the true Pole until the year 2095 when it will be within 1 1/2 minutes from it. Then it will continue moving slowly in the great precessional circle until in nearly 26,000 years it will again occupy the same approximate position. See also Pole Star.

polarization.   A preferential orientation of the waves making up a light beam so that the waves lie in a single plane. Light may be polarized by being reflected from or passing through certain materials. If the direction of the vibrations or waves remains constant along the light beam, so that the vibrations are all in the same plane, we say the light is plane polarized. If the direction of the electrial maximum is rotating uniformly in space and tracing out a helical path something like the thread of a screw, the light is circularly polarized. If the motion is spiral but the intensity varies during the cycle, the polarization is elliptical. Astronomers utilize polarization in the interpretation of the physical condition of planets, stars, nebulae, and even interstellar space.

polarimeter. A device for measuring the amount of polarization in a light beam (i.e., the degree to which the beam is polarized expressed as a percentage of polarization).

pole. A point on a rotating body where its axis of rotation cuts the surface.

poles, celestial. (1) Intersections of a body's rotation axis extended to the celestial sphere. (2) Specifically, Earth's celestial poles.

pole of the galaxy. The two points 90° from the plane of the Milky Way.

Pole Star. Any fairly bright binary star close to the North or south pole of the celestial sphere. Today, Polaris is the North Pole Star and there is no southern one. Because of precession, the celestial poles describe a large circle in the sky. Thus in about 2900 B.C. Alpha Draconis was the North Pole Star. Future candidates for the position lie in the constellations Cepheus and Lyra.

polishing. Smoothing the surface of a mirror or lens with a very fine polishing compound and a soft lap (pitch) as opposed to grinding it, which involves altering the curvature with a coarse abrasive and a metal or glass disk.

Pollux. See Beta Geminorum.

polyconic projection. A projection that is made on a series of cones tangent to the Earth. Meridians, except the central one, are represented as curved lines and parallels of latitude are nonconcentric circles. There is no distortion on the central meridian and (unless the spread in longitude is great) distortion elsewhere is small.

polytrope. A theoretical mathematical model of a compressible gas sphere (like a star) in equilibrium under its own gravity. Its density increases inward with pressure, and various density-pressure relations are defined by choice of the so-called polytropic index. Polytropes are useful approximations used in theoretical work on stellar interiors.

population. A group of stars and interstellar material with similar compositions. Two main populations exist with many stars classed in intermediate groups. See p. 182.

Population I stars.   Stars, gas, and dust found in the neighborhood of the Sun and believed to exist in the disks of all spiral galaxies, but not "above" and "below" the disks in the halos of these galaxies. These stars and associated material have relatively high content (a few percent) of elements heavier than helium, such as metals. The brightest Population I stars are massive blue stars only recently formed. Population I is associated with regions where star formation continues to the present day.

Population II stars.   Stars, gas and small amounts of dust found in globular clusters, in elliptical galaxies, and in the central regions of spiral galaxies. Population II stars are older than Population I stars, have very low content of elements heavier than helium, and are found in the nuclei and halos of spiral galaxies and in elliptical galaxies. The most luminous Population II stars are the relatively old (evolved) red supergiants. Population II stars are believed to be the first stars formed when galaxies began to accumulate from nearly pure hydrogen-helium gas following the "big bang" that initiated the known universe.

position angle (PA).   For an elongated object (or binary star), the angle (measured counterclockwise) from the north direction to the axis of the object (or to a line from the brightest star to the faintest). A line drawn due north.

positron (anti-electron).   A subatomic particle which is identical to the electron in atomic mass, theoretical rest mass, spin and energy, but is positively charged. The positron is short lived.

potassium-argon date.   Date of a rock, such as a meteorite, measuring the time since solidification (or more specifically, since argon gas, a daughter product of radio-activity, has been retained in the rock minerals). In some meteorites it dates the time since the last major collision of the parent body.

potential temperature.   The temperature which a sample of air would assume if lowered (or raised) along a dry adiabat to a standard pressure of 1000 millibars. No condensation may occur.

power.   The magnification used when observing with a telescope. It is given by the focal length of the telescope objective divided by the focal length of the eyepiece (f). An astronomical telescope is equipped with several eyepieces of different focal length so that the magnifying power may be changed at will. For example, a telescope with a focal length of 200 cm and a 2 cm eyepiece gives a power of 100x.

power of ten notation.   In very large and very small numbers (such as 10 billion or 1 millionth) it is inconvenient to write out the many zeros. Since 10 squared, 10^2, is 100 (the number 1 followed by 2 zeros), and 10 cubed, 10^3, is 1000 (1 followed by 3 zeros), etc., that is, since the superscript, or "power" is equal to the number of zeros, we can write 10 billion (1 followed by 10 zeros) as 10^{10}. Similarly, 0.01 is 1/100, or 10^{-2}, and 1 millionth is 10^{-6}. A number such as 3 billion is written 3×10^9; and 4.35×10^{-5} means 0.0000435.

Poynting-Robertson Effect.   A spiraling of small orbiting bodies inward toward the Sun or central body because of the influence of solar (or stellar) radiation. For bodies of appreciable size, the influence of the radiation is insignificant, but for small particles, the situation is different. In addition to the normal force directed away from the Sun due to solar radiation pressure, the particle absorbs radiative energy primarily from the leading illuminated side. The illuminating photons transfer momentum, or force, from this direction. Since the body reradiates this energy in all directions, the force is not balanced and must be added to gravity. Calculations show that, as a result, a drag acts on the particle and decreases its orbital momentum. Consequently, the particle follows a spiral path, drawing steadily closer to the Sun or other central body. The magnitude of the effect is inversely related to the radius and density of the particle, and is important in solar-system particles at sizes of about a centimeter or less.

Praesepe (The Beehive).   A very prominent open star cluster, also known as M44 or NGC 2632. It is clearly visible to the naked eye whenever the sky is clear and dark, and is resolved into many stars in a telescope. At the point halfway between Regulus and Pollus, two faint stars of Cancer can be made out—Delta (4.2) and Gamma (4.7). Between the two, slightly west of a line joining them will be seen the faint shimmer of Praesepe, aptly nicknamed The Beehive. It is estimated to be 400 million years old, 520 light-years away, and 13 light-years across.

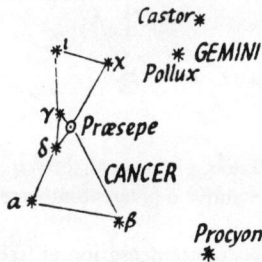

precession.   (1) A slow change in the direction in space of the Earth's axis of rotation, due to gravitational action of the Sun and Moon on the oblate Earth. (2) The slow conical motion of the rotation axis due to external torque acting on the rotating body. (3) Any similar shift in orientation of a planet or star due to gravitational forces of nearby bodies. See also general precession.

precession in declination.   The component of general precession along a celestial meridian (about 20 seconds of arc per year).

precession in right ascension.   The component of general precession along the celestial equator (about 46 seconds of arc per year).

precession of the equinoxes.   A slow shift in position of celestial coordinates relative to the stars. The ecliptic is the great circle where the plane of the Earth's orbit cuts the celestial sphere. The celestial equator is another great circle where the plane of the

Earth's equator cuts the celestial sphere. These two circles are inclined to one another by 23.5°. The two points of intersection are known as the equinoctial points, because when the Sun reaches them, (on about March 21 and September 21) days and nights are equal. It is known today that Earth's pole moves along a circle of radius 23.5° in a period of some 26,000 years. This phenomenon, known as the precession of the equinoxes, causes the equinoctial points to shift westward (clockwise when one looks down from the north) along the ecliptic at a rate of some 50.2 seconds of arc per year. All right ascension and declination coordinates, which are tied to the celestial equator likewise change from year to year in a 26,000-year cycle. Precession has the interesting side effect of causing the Sun to be located far from the constellations associated with its position in astrology columns, which follow the astrological practice of about 3,000 years ago, when the associations of Sun and constellations were formulated.

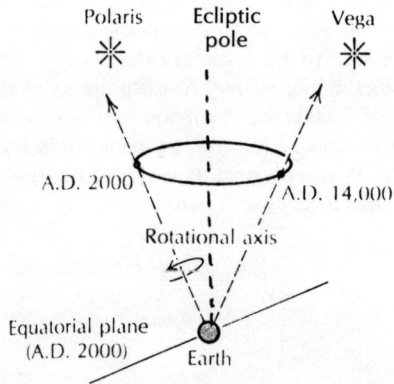

precipitable water. The thickness of the layer of water that would result if all the water vapor in the atmosphere above a given point were condensed at a given time.

precipitation. Falling products of condensation or freezing such as rain, snow, hail, drizzle. Precipitation elements are usually larger than 10^{-2} cm. Particles smaller than this size usually remain supported in the air as clouds.

precede. To lie west of, in the sky, so as to cross the meridian prior to a reference object.

pressure. The force applied over a surface divided by its area. Expressed in units of force per unit area (as dynes per square centimeter or pounds per square inch).

pressure broadening. An increase in wavelength span (broadening) of a spectral line due to pressure in the source gas.

pressure wave. In meteorology, a short period oscillation of pressure such as that associated with the propagation of sound through the atmosphere; a type of longitudinal wave.

primary. (1) The most massive body in any system of bodies revolving under gravitational forces about their common center of gravity. (2) The brightest star in a binary or multiple star system (often but not always agreeing with definition 1).

primary minimum. The deepest minimum in the light curve of an eclipsing binary.

primary mirror. The largest mirror in a reflecting telescope.

prime focus. The point at which the rays from the primary mirror of a reflecting telescope come to a focus.

prime focus photography. Celestial photographs in which the film is placed at the prime focus of the telescope.

prime meridian. The zero-longitude meridian on a planet's surface from which longitude is measured (on Earth, the meridian of Greenwich, England).

primeval atom. Term used by Lemaitre (father of the big-bang theory) to refer to the hypothetical mass of crammed-together nuclear particles that exploded in the big bang forming the matter now observed in the universe.

primitive. Little-altered since the period of origin (usually in reference to solar system origin, 4.6×10^9 years ago, as in "primitive meteorite.")

prism. In optical work, a transparent substance bounded by plane-polished surfaces inclined to one another. A prism produces refraction as light waves pass through it, spreading the beam in to an array of different colors, or spectrum. It is therefore often the heart of spectrographs or related devices.

probable error (pe). A statistical measurement of the certainty with which a value is known. The probability is 50 percent that the true value will turn out to be more than one pe away from the measured value, according to the statistical definition. See also standard error.

Procyon. See Alpha Canis Minoris.

profile. The plot of intensity versus wavelength across a spectral line. From the line profile, motions and other properties of the gas producing the line can be inferred.

prograde. Implying rotary motion from west to east around a center.

prominences. Clouds of hot, luminous gas, best seen when they project beyond the edge of the Sun's disk. They are structures in the solar atmosphere, elevated above the surface of the photosphere, usually in continuous and often rapid movement. The gas in the prominence, at about $10^{4\circ}$ K, is cooler than the surrounding million-degree gas at the lower corona.

proper motion.   Angular rate of motion of a star (seconds of arc per year) across the sky, corresponding to stellar motion at right angles to the line of sight, relative to the solar system.

proton.   One of the major nuclear particles, with positive charge and mass equaling 1.67×10^{-24} grams, or 1836 times the mass of an electron. The atomic nucleus of each atom is made up of a number of two types of particles, called protons and neutrons. These two particles have approximately the same mass and unity on the conventional atomic weight scale, but they differ in that the proton has a unit positive charge of electricity, whereas the neutron is uncharged electrically, and is neutral. Because of the protons present in the nucleus the nucleus has a positive electrical charge.

proton–proton chain.   The series of nuclear reactions providing the main source of energy in the Sun. Although the overall nuclear reaction in the Sun can be represented by the combination of four hydrogen atoms to yield a helium atom, the probability that four protons will combine in a single step is small. The process actually takes place in stages. At temperatures below about 15 million°K, the proton–proton chain describes the main stages. In the proton–proton chain, the first step is the combination of two protons to form a deuteron, i.e., a nucleus of deuterium, the heavier isotope of hydrogen, and a positron. The deuteron then combines with another proton to form helium-3, a helium nucleus with two protons and a neutron. Third, two helium-3 nuclei combine to form two nuclei with two protons and two neutrons.

proton storm.   The cloud of protons shot into space by a solar flare.

protoplanet.   (1) Any of the Sun's planets as it emerged or existed in the formative period of the solar system. (2) In certain theories, a gravitationally collapsed primeval form of one of the planets, equaling or exceeding the planet's current mass.

protostar.   A star during its formative stages, generally prior to the onset of nuclear reactions in its interior. A protostar is generally viewed as a contracting gas cloud.

protosun.   The Sun during its formation and prior to or during the early formation of the planets.

Proxima Centauri.   The nearest star to Earth (4.3 light-years), of magnitude 11, in the constellation Centaurus. It is a member of a multiple system containing the bright star Alpha Centauri.

Ptolemaic theory (or system).   A theoretical (and incorrect) system for explaining the apparent motions of the Sun, Moon and planets developed or extended by Claudius Ptolemaeus in the 2nd century A.D. The planets were supposed to move in the circumference of small circles (epicycles), the centers of which revolved around the Earth. Below is the Ptolemaic System of planetary motions. (See Fig., p. 187.)

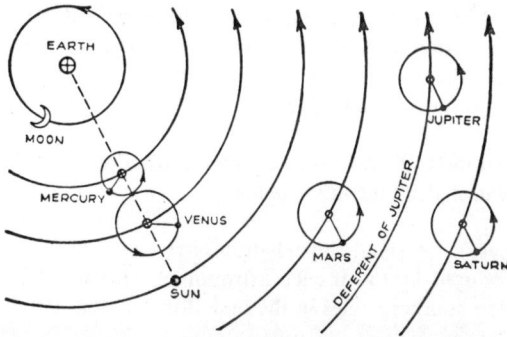

pulsars.   As defined observationally, variable radio stars which have very short (around a second) and extremely stable periods of pulsations. These objects were first detected late in 1967 by Jocelyn Bell and A. Hewish, of the Mullard Radio Astronomy Observatory, Cambridge, England. During the course of their radio observations they noted a celestial source that "pulsed" regularly every 1.3 seconds. Others were later found. Pulsars are believed to be very rapidly rotating neutron stars (q.v.) which are remnants of supernovae explosions. Electrons moving rapidly in their magnetic field produced narrow beams of radiation which sweep around the sky as the pulsar spins with its (roughly) one-second periods, rather like search light beams. Hence we see the flashing of these beams. Following discovery of the radio pulsing, flashing of the visible light of these objects with the same period has also been confirmed.

pulsating variable.   A variable star with brightness changes due primarily to an alternative expansion and contraction of the star (as opposed to other variables such as eclipsing binaries).

Puppis (The Stern).   One of the constellations formed by Gould in 1877 from the ancient, obsolete constellation Argo Navis. It has an area of 673 sq deg. The Milky Way is rich in this region and sown with bright scattered stars so that Puppis offers a fine selection of objects for the observer. This constellation includes at least 40 open star clusters. There is one globular cluster NGC 2298, six planetary nebulae, and the pulsar Puppis A.

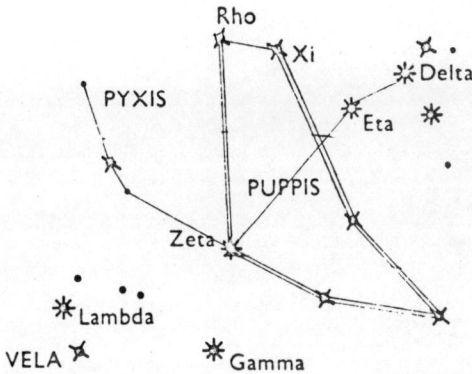

Purkinje Effect.   The more rapid fall-off of the eye's red sensitivity than blue sensitivity as a light source is dimmed. If we have two point sources of light, one of which is red and the other white, both of the same brightness, and then reduce each by the same amount, it is found that the red source will appear to be the fainter of the two.

pyrheliometer.   An instrument for measuring the total intensity of solar radiation, both direct and scattered by the atmosphere.

Pyxis (The Compass).   A small constellation once a part of the now obsolete ship Argo. It was introduced by the French astronomer Nicolas de Lacaille during his pioneer work on the southern skies in the mid-19th century.

Q

quadrant. A form of sextant with a graduated run forming roughly one-quarter of a circle.

quadrature. The position of a superior planet or other body normal to the Earth-Sun direction as seen from the Earth. An elongation of 90°, usually specified as east or west in accordance with the direction of the body from the Sun. The Moon is at quadrature at first and last quarters.

quantized. Divided into small indivisible units; not continuous.

quantum. (1) The minimum discrete amount of any quantity. (2) The smallest amount of energy that can be radiated or absorbed by matter. Each quantum of radiated energy, or photon, has a certain wavelength, and the amount of energy in the photon is inversely proportional to its wavelength.

quantum efficiency. Probability that a photometric instrument will detect a single specific photon entering the instruments (i.e., efficiency at detecting photons).

quarks. Hypothetical elementary particles having fractional electric charges (i.e., a fraction of an electron's or proton's charges).

quasars (or QSO). Acronym for quasi-stellar radio source (or quasi-stellar object). In 1960 astronomers first discovered that many previously unidentified radio sources were associated with what appeared to be unresolved stellar images. These were first called quasi-stellar radio sources, later shortened to "quasars." They appear to exist throughout space and show unexpected properties. They are very small (estimated from their short periods of variability to be only a few light-years across and less) and extremely bright, the brightest having almost 100 times the luminosity of the brightest known galaxies. They emit tremendous amounts of both optical and radio radiation and show spectra that are somewhat like those of the nuclei of Seyfert galaxies, but very highly red-shifted. Many astronomers believe they are nuclei of distant galaxies with exploding centers; others believe they are smaller and nearer.

quiet sun. The Sun during periods of minimum radio and sunspot activity, which occur twice during the 22 year solar sunspot cycle.

R

r-process. The buildup of heavy elements inside a massive star or supernova by means of very rapid (hence "r") capture of neutrons by atomic nuclei, thus synthesizing heavy nuclei.

rad. The quantity of any ionizing radiation that leads to the absorption of 100 ergs of energy per gram of irradiated material.

radar. A short-wave radio method of determining distance and direction by beaming a radio pulse toward the object from a station on Earth; the pulse is reflected back from the object and the time is observed when the return pulse is received at the station. Since radar waves travel with the speed of light, 300×10^8 meters per second, the distance is directly determined by the delay time.

radar astronomy. The study of celestial bodies within the solar system by means of radiation originating on Earth but reflected from the body under observation.

radial velocity. A star's relative velocity directly toward or away from us. The radial velocity is measured by the Doppler shift of the lines of its spectrum from their normal positions. By the Doppler effect the wavelengths are shortened (made more blue) if the star is approaching or reddened if receeding.

radial velocity curve. The curve obtained by plotting the radial velocity against time for a binary star. The curve reveals the alternate approach and recession of the star as it orbits around its companion.

radian. An angular unit equaling $57°.3$ or $206,265''$, defined as the angle which has an arc length along a circle equal to the circle's radius.

radiant. Point on the celestial sphere from which meteors of a given shower appear to radiate. The effect is due to perspective.

radiation. (1) Energy propagating as electromagnetic waves, i.e., (in order of increasing wavelengths) gamma rays, x-rays, ultraviolet light, visible light, infrared light, radio waves. (2) (Occasional usage) Energy propagated in the form of fast moving atomic particles, such as protons and electrons, as in "cosmic-ray radiation."

radiation belt. A toroidal envelope of charged particles trapped in the magnetic field of a planet. Earth and Jupiter have such belts, called Van Allen belts.

radiation, coherence. See coherent radiation.

radiation intensity.   The energy received from a unit solid angle per unit area in unit time.

radiation laws.   Relations between the temperature of a body and the quantity or quality, or both, of the radiation it emits. See Kirchhoff's Laws, Planck's Law, Stefan's Law, and Wien's law.

radiation pressure.   Pressure exerted by light or other radiation upon an object due to the delivery of momentum to the object by the photons that strike it. It is negligible for large, massive objects like planets, but can drive interplanetary microscopic particles out of the solar system.

radio astronomy.   Astronomical research using electromagnetic radiation at radio, rather than visible, wavelengths. It has revealed much about radio-emitting stars, radio galaxies, and hydrogen in space. Neutral hydrogen atoms in space can emit radio waves of wavelength 21 cm. Since hydrogen is abundant in space, the 21-cm line is one of the most important in radio astronomy.

radio galaxies.   Galaxies that emit measurable amounts of radio radiation. At least a hundred such galaxies have been identified, being generally classified into two broad categories, called normal galaxies and peculiar galaxies, without a clear line of demarcation between them. Normal radio galaxies are mostly spirals since they are the most common galaxies observed, but there are also elliptical and irregular galaxies in this category. The essential characteristic of a normal radio galaxy is not its optical appearance, but the fact that its radio emission is regarded as normal. The rate of radio energy emission by a peculiar radio galaxy is larger by a factor of a hundred to a million or so than the emission from a normal galaxy. The galaxies that are peculiar in both radio and optical respects are those which are evidently single galaxies, often with evidence of explosive activity in their centers, such as a "jet" extending from the nucleus. A few others appear to involve two (or perhaps more) interacting or colliding galaxies.

radio meteor.   A meteor detected by the reflection of a radio signal from the trail of ions that it leaves as it passes through the atmosphere.

radio stars.   Stars detected by radio waves emitted by them. In the nearly two thousand radio sources few correspond with prominent visual stars; many star-like radio sources have turned out to be distant galaxies. Others are supernova remnants or other objects not prominent in visual light.

radio sun.   The Sun as revealed by a radio telescope, with a diameter about twice that of the visible Sun. The radio waves emitted from the Sun have relatively short wavelength.

radio telescope.   A large highly directive radio receiver for detecting radio waves emitted by celestial bodies or by space vehicles.

radio window.   Radio wavelength intervals in which radio waves are not blocked by the atmosphere. Absorption of radiation by Earth's atmosphere takes place over wide ranges of wavelengths but there are two major "windows" through which celestial objects can be observed: the optical and radio windows. The radio window extends from microwaves to wavelengths of about 15 to 30 meters (20 to 10 megacycles per second) and sometimes beyond, through which observations can be made from Earth. The long wavelength (low frequency) limit depends upon the condition of the ionosphere.

radioactive.   Referring to unstable atoms that spontaneously break apart into smaller atoms, often emitting subatomic particles at the same time. The second-generation atoms are called daughter atoms, or daughter isotopes.

radiometer.   A radio receiver designed to measure all thermal (infrared) radiated energy received per second from a target object.

radioisotopes.   Isotopes that are unstable and thus radioactive. Most are produced by various nuclear reactions, in particular by fission. A characteristic of the instability (or decay) of radioisotopes is that the nuclei emit electrically charged particles, often a positron or electron. As a consequence of particle emission, the original, or parent, nucleus is converted into a "daughter" nucleus of a different element. This daughter nucleus may be stable or it in turn may be a radioisotope and emit another particle, and so on.

radiosonde.   An instrument carried aloft by a balloon or an airplane which broadcasts, by means of a miniature radio transmitting set, the atmospheric temperature, pressure, and humidity encountered.

radius vector.   A straight line connecting a fixed reference point or center with a second point, which may be moving; specifically in astronomy, the straight line connecting the center of a celestial body with the center of a body which revolves around it, as the radius vector from Sun to Earth.

rainbow.   An optical refraction phenomenon consisting of a circular arc of radius about 41° composed of alternating bands of color (violet inside, red outside) seen in the sky during rainstorms, when the clouds break sufficiently to allow the Sun to illuminate falling rain. Occasionally, an outer rainbow of radius about 53° can be seen with reversed color sequence (red inside, violet outside). Rainbows of light from the Sun as it passes through spherical water droplets. The optics were first successfully explained by M. A. deDominis in 1611.

Raman scattering.   Redirection (scattering) of light by small particles, involving changes in color as a result of interactions between photons and atoms or molecules.

Ramsden disk.   The cross section of the exit pupil at the point of the sharpest image.

Ramsden eyepiece.   An eyepiece made up of two plano-convex lenses with convex surfaces facing each other. The field of view is in the neighborhood of 35° to 40°, although not all of this wide field can be used effectively since there is considerable chromatic difference in magnification. Objects near the edge of the field are surrounded with reddish fringes.

random walk.   Motion of any particle (usually microscopic) by means of discontinuous movements from the site of one collision (or interaction with another particle) to the site of another. Motions on the intervening distances, called the free path, may be much faster than the net rate of progress. Typical particles performing random walk motions are dust particles floating in a gas and photons moving inside stars.

ray.   (1) A single line or narrow beam of light. (2) (Now rare) A beam of atomic particles, such as electrons. (3) A streak of bright material ejected from a fresh-looking crater and lying across the adjacent surface of the Moon or other planetary body. (Rarely, rays may be darker than the background material.)

Rayleigh atmosphere.   An idealized atmosphere consisting of only those particles, such as molecules, that are smaller than about one-tenth the wavelength of most radiation incident upon that atmosphere.

Rayleigh scattering.   Redistribution (scattering) of light beams in different directions by gas or dust particles much smaller than the light wavelength. In Rayleigh scattering, the short waves (blue light) are scattered much more than the long waves (red) which pass on through. Rayleigh scattering accounts for the sky's blue color.

Rayleigh standard.   The requirement that all reflected rays from a mirror have no deviation greater than $1/4$ wavelength of light.

R Coronae Borealis Variables.   A small and peculiar class of variable star typified by R Coronae Borealis. Normally this star is just visible to the naked eye but at irregular intervals it fades, often quite rapidly to anything between 7th and 14th magnitudes. The eventual climb back to maximum is more protracted and often accompanied by marked fluctuations. These stars vary "in reverse," spending most of their time at maximum and suddenly dropping by up to 6 or 7 magnitudes to a maximum that can last for weeks or months.

recombination.   Recapture of an electron by a positive ion which previously lost an electron.

recurrent nova.   A star in which the nova outburst occurs at fairly frequent intervals of the order of tens of years, the amplitude in general being smaller than in ordinary novae.

reddening.   See space reddening.

red giant.   A large low temperature star having high luminosity. The red giants have surface temperatures that range from about 2000°–3500° K and so are orange or red in color. Because of their large size, they are from a hundred to several thousand times more luminous than main-sequence stars of the same spectral class. They represent a late stage in the evolution of most stars occurring after hydrogen has been consumed in the inner regions of the star.

red-shift. (1) Any displacement of spectral lines toward the red, particularly in distant galaxies. An examination of the spectra of a number of distant galaxies, begun by V. M. Slipher in the United States in 1912 and culminating about 1926, showed that the characteristic lines, such as the H and K lines of ionized calcium, all appeared at longer wavelengths than in the spectrum of the Sun and of other stars in the local galaxy. The change in wavelength toward the red end of the spectrum which has been observed for several hundred galaxies, has become known as the red-shift. The red-shift is believed to be a Doppler shift (q.v.) indicating that these galaxies are receding from us. The speed of recession of the most distant cluster of ordinary galaxies thus far observed is 108,000 km/sec. However, in addition to Doppler-caused red-shifts, there are gravitational red-shifts. As a consequence of the general relativity theory, the periods of identical oscillations depend on the gravitational potentials at the point where they are situated; so that for example, the wavelength of a spectral line coming from the Sun exceeds the corresponding line on the Earth by the fraction 2.2×10^{-6}. Doppler and gravitational red-shifts cannot be distinguished by measurement of the spectral-line positions, giving rise to controversy about causes of red-shifts among some objects. In particular, quasars have extremely high red-shifts, interpreted as Doppler shifts by many astronomers, but as non-Doppler shifts by some others. See also relativistic red shift. (2) A quantitative measure of red-shift designated as $z = shift\ in\ wavelength/original\ wavelength$.

red spot. The only semipermanent feature seen upon Jupiter. It was a red colored ellipse 40,000 km long and 11,000 km wide when very prominent in 1878, but has since undergone changes from greater to lesser prominence, sometimes fading to a more gray color. It is an atmospheric disturbance of unknown cause. It was reported shrinking somewhat in 1979. Also called Great Red Spot.

reddening. Reddening of starlight caused by interstellar scattering in dust clouds. It arises as blue light is scattered out of the beam, while red light passes on through. A view of the star reveals, therefore, primarily its red light and not its blue light.

reflecting telescope. See reflector.

reflection. The return of waves of light or sound after striking surfaces or diffuse material.

reflection coefficient. A measure of the proportion of light reflected by a given surface; defined as the ratio of the radiant energy reflected along the geometrical reflection path to the total that is incident upon the surface. By definition a reflection coefficient of 1.0 implies perfect specular reflection.

reflection effect. An irregularity in the light curve of an eclipsing variable due to the hemisphere of the fainter star which faces the brighter companion being brighter than the other.

reflection nebula.   A nebula whose light is mainly light from an adjacent star reflected off nebular dust grains, as opposed to light emitted from the atoms of gas in the nebula itself.

reflectivity.   A measure of the fraction of radiation reflected by a given surface; defined as the ratio of the radiant energy reflected to the total that is incident upon that surface. Loosely (but improperly) used to indicate reflection coefficient.

reflector.   A telescope using a mirror to collect and reflect light to a focus. The simplest type is the Newtonian which can be constructed in almost endless variations from the short-focus telescope designed for wide-field observing to the long-focus giants used in studying remote galaxies. The Newtonian, regardless of its aperture, is a perfect achromatic telescope, since all colors focus at the same point. In apertures of about 18 cm (7 inches), reflectors are superior in light-gathering power to refractors of equal size. Below is a diagram of the types of optical systems used in reflecting telescopes: (a) prime focus; (b) Newtonian focus; (c) Cassegrainian focus.

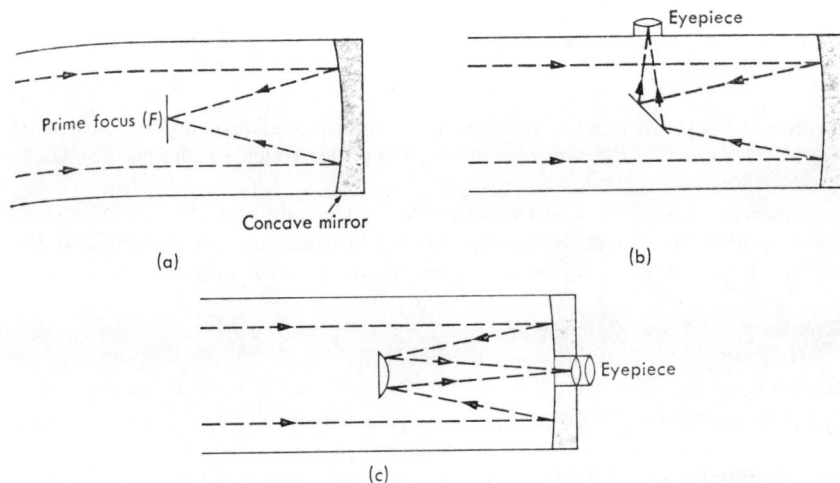

refraction.   The process in which the direction of a beam of radiation is changed as the result of a change in density within the propagating media, or as the energy passes through the interface representing a density discontinuity between two media. In the first instance the rays undergo a smooth bending over a finite distance. In the second case (as when light enters a lens) the index of refraction changes through an interfacial layer that is thin compared to the wavelength of the radiation; thus the refraction is abrupt and essentially discontinuous.

refraction of the atmosphere.   The bending of a ray of light or a radio wave from a star in passing through the Earth's atmosphere due to the decrease of density from the surface upwards. The amount of refraction of light varies from zero at the zenith to one minute of arc at altitudes of about 45°, and about 35 minutes of arc with the celestial body on the horizon; the body appears higher above the horizon than with no atmosphere.

refractor. A telescope in which a large lens called the objective lens forms images of distant objects. The objective generally contains two or more glass pieces cemented or separated in a design that reduces color aberrations. An astronomical refractor gives an inverted image and requires an erecting eyepiece or prism system. The Galilean refractor or "spyglass" gives an erect image, seldom has a magnifying power greater than 5, and usually has an aperture of less than 2 inches. Its great deficiencies as an astronomical instrument are its low resolving power, its small field, and its lack of image brightness. Below is the principle of a refracting telescope.

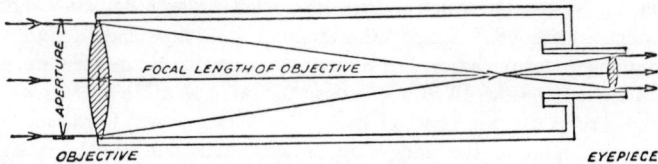

APERTURE

FOCAL LENGTH OF OBJECTIVE

OBJECTIVE   EYEPIECE

refractories. Elements or chemical compounds with high melting temperatures, which tend to be left behind during heating of planetary matter and escape of its volatiles (q.v.).

regolith. A layer of fragments and powdery solid produced by impacts of meteorites on the surface of the Moon or other satellites, asteroids, or planets.

regression of the nodes. A slow westward drift of the nodes of the Moon's orbit, completing a 360° drift along the celestial equator in 18.6 years. The movement of the nodes is by perturbations of the lunar orbit by the Sun and Earth.

Regulus (Leo). See Alpha Leonis.

relative humidity. The ratio of the actual vapor pressure to the saturated vapor pressure. Usually designated by f.

relative movement. Motion of one object or body measured relative to another. Usually called apparent motion when applied to the change of position of a celestial body as observed from the Earth.

relative position. A point defined with reference to another position, either fixed or moving.

relative visibility factor. The ratio of the apparent brightness of a monochromatic source to that of a source of wavelength 5500°A having the same energy.

relativistic. Pertaining to material or particles moving at speeds which are an appreciable fraction of the speed of light. Properties of relativistic particles such as mass differ from those measured at rest as specified by Newtonian physics for lower speeds.

relativistic red shift.   A red spectral shift due not to the Doppler effect, but to effects predicted in the theory of relativity, such as high gravitational fields.

relativistic velocity.   A velocity equaling an appreciable fraction (say, more than 1/10) of the speed of light.

relativity.   (1) The principle that motion of a particle can be defined or measured only relative to another specified particle or coordinate system, and that no motion can be determined to be "absolute," or definable without reference to such particles or coordinates. (2) Loosely, Einstein's theories of special and general relativity.

relaxation.   The process of a system coming back to equilibrium after that system is disturbed.

residual.   In celestial mechanics and trajectory analysis, the deviation between an observed and a computed value, usually observed value minus computed value.

resolution.   (1) The ability of a film, a lens, a combination of both, or a vidicon system to render fine detail, in particular a standard pattern of black and white lines. (2) The dimension (usually angular; sometimes equivalent linear) of the smallest detail visible in a given optical or photo system.

resolving power (telescope).   The ability of a telescope to separate small angular objects such as close pairs of stars, or craters on the Moon. It may be given in seconds of arc, approximately by 5.4 divided by telescope aperture in inches. Thus a 5 inch telescope would have resolution about 1 second of arc.

resonance.   Enhanced response of any system (mechanical, orbital, electromagnetic) when stimulated from the outside by a periodic disturbance having the same frequency as the system's own natural frequency of oscillation.

reticule.   A network of fine lines, wires or crosshairs placed in the focus of the objective of a telescope or other optical instrument in order to give reference lines for orienting or measuring the image.

Reticulum (The Net).   A constellation defined in 1752 when Lacaille renamed an earlier group. It is a small constellation of area 114 sq degrees near the large Magellanic Cloud. There are several conspicuous nebulae in Reticulum, no clusters or galactic nebulae occur.

retrograde motion.   Motion in an orbit opposite to the usual orbit direction of solar-system bodies. Motion from east to west around a center. Refers also to apparent motion on the celestial sphere.

reversing layer (outdated).   A gaseous layer between the photosphere and the chromosphere of the Sun regarded responsible for the Fraunhofer absorption lines in the solar spectrum. The term is a simplification, since the lines actually arise throughout the photospheric gas.

revolution.   (1) Orbital motion of a body around its primary. (2) Any motion around a center outside the body. Contrasts with rotation.

Rhea.   A satellite of Saturn orbiting at a mean distance of 527,000 km. Its diameter is estimated at about 1600 km.

rhumb line.   A line which makes equal angles with all the meridians it crosses.

ridge.   An elongated atmospheric area of relatively high pressure.

rift.   (1) An apparent division of the Milky Way caused by dark clouds of superposed dust. (2) A large-scale fracture in a planetary crust caused by motions of adjacent crustal land masses away from each other. The Red Sea, Gulf of California, African Rift Valleys on Earth and Valles Marineris on Mars are examples.

Rigel.   See Beta Orionis.

Rigel Kent.   See Alpha Centauri.

right ascension (RA).   Angle measured eastward from the vernal equinox to the foot of a star's hour circle (i.e., projection of Earth's longitude lines upon the sky). Right ascension and declination are the astronomer's usual coordinates for locating positions in the sky—analogous to longitude and latitude on the surface of the Earth. Below is a diagram showing right ascension and declination, altitude, and azimuth, and their relation for a particular star. The observer is situated in the center of the sphere.

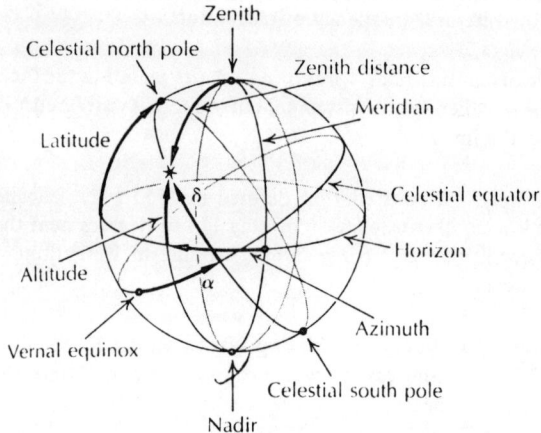

right ascension circle.   A circle placed on the polar axis of a telescope to aid in lining up the instrument with the right ascension of a celestial object.

rilles (moon).   Lunar valleys, typically a kilometer or two across and many kilometers long. Some are straight and some are winding. Their origin or cause is not clear. Straight examples may be related to graben; others may be slumps into lava tubes or eroded channels cut by flowing lava. Apollo 15 astronauts found layered rock outcrops (presumably lava flows) exposed through rubble in the walls of the Nadler rille.

rime.   A white, opaque, granular structure consisting of very small ice particles which have little cohesion.

ring galaxy.   A galaxy with a prominent bright ring surrounding the center, which in some cases is faint.

ring plain.   Obsolete term for large lunar craters.

Ring Nebula (NGC 6720).   Between Beta and Gamma Lyrae, but not visible to the unaided eye, is Ring Nebula, one of the most interesting of the planetary nebulae. It is in the form of an ellipse, in the center of which is a very faint star, surrounded by a mass of nebulous matter. Its distance is estimated to be 2300 light-years and its diameter about 0.7 parsecs. It is probably the expanding shell of gas thrown off a supernova.

ring stars.   Stars whose spectra indicate they are surrounded by a ring or disk of gas.

R Leonis.   A red long-period variable with a period of 312 days; though like all its kind, it is not perfectly regular. At maximum it may attain magnitude 5, but at minimum it drops below 10. Although seldom visible with the naked eye, it is not hard to locate when at its best, since it lies conveniently close to Regulus.

Roche Limit.   A theoretical critical distance of a zero-strength (e.g., liquid) satellite from its primary, within which it would be shattered to fragments by the tidal forces. The distance is 2.44 radii of the primary from its center (assuming equal density for satellite and primary). Saturn's ring system is largely within the limit. Satellites with non-zero strength, for example ice or rock bodies, would have to be still closer to the primary to be fragmented by these forces.

Roche lobes.   A figure 8-shaped volume surrounding a circular-orbiting binary star pair, the size and shape of which depends upon the masses of the stars and any external gravitational fields. Within this equipotential surface, material with less than a specified energy will not escape but circulate between the two stars, probably eventually colliding with one or the other.

rocket.   A vehicle that accelerates by expelling part of its own mass (i.e., fuel) at high velocity out of a rear nozzle.

Roentgen ray.   Obsolete term for x-ray.

Ronchi test.   A derivative of the Foucault test for telescope mirrors in which the knife edge is replaced by a screen or grating.

rotation.   Turning of an object on its axis. Contrasts with revolution.

rouge.   Form of iron oxide used in polishing optical surfaces.

RR Lyrae variable stars.   Named after the variable star RR Lyrae. They are often called cluster variables and are recognized in great numbers inside, as well as outside, globular clusters. The periods of their light variations are around a half a day, ranging from 1 1/2 hours to about a day, sometimes slowly changing. They are related to Cepheid variables and are Population II stars. Their known luminosities allow them to be used as distance indicators.

rubidium-strontium date.   Date of solidification of a rock sample based on measurement of minute quantities of radioactive and radiogenic atoms in the sample. It is a highly successful method of dating lunar and ancient terrestrial rocks, as well as meteorites.

runaway star.   Stars with very high space velocities, often moving away from a region of recent star formation. They may be escaped members of binary systems arising when the more massive members undergo a supernova explosion. Most are O or B stars.

RV Tauri variable stars.   These stars have irregular periods, usually of several months. The minimum magnitude is usually fairly constant, but maxima are alternately bright and faint. The range is usually 2 or 3 magnitudes, and the period quoted is for the whole cycle of bright and faint maxima. R Scuit is a bright RV Tauri type, sometimes reaching magnitude 5.

RW Aurigae variable stars.   See T Tauri variable stars.

S

S-process. Buildup of heavy elements inside massive stars by means of slow (hence "S") capture of neutrons by atomic nuclei, thus synthesizing heavy nuclei.

Sagitta (The Arrow). A small northern constellation entirely immersed in the Milky Way between Aquila and Vulpecula. Ptolemy assigned five stars to it, these make an elongated group 10° north of Altair in Aquila. The area is only 80° sq degrees.

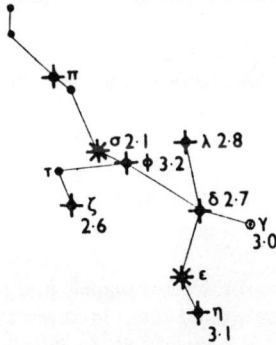

Sagittarius. A major southern constellation containing many clusters and the region of the galactic center.

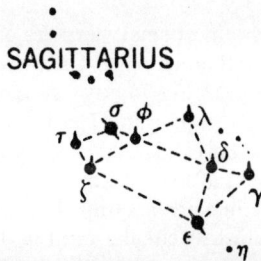

SAGITTARIUS

Sagittarius A. The radio source marking the galactic center, a region about 12 parsecs (39 light-years) across.

Sagittarius arm. The spiral arm of the Milky Way, adjacent to our spiral arm but closer to the center.

saros. The interval of 18^y 11 $1/3^d$ (or a day less or more, depending on the number of leap years included) after which a series of eclipses is repeated. It was thus useful in ancient times for predicting eclipses once earlier series of eclipses were known. It equals

223 synodic months (6585.32 days) and is nearly the same length as 19 eclipse years (6585.78 days). After a Saros interval, the Sun and Moon have returned to nearly the same position relative to each other and to the node, and their distances from us are nearly the same as before, allowing recurrence of a similar pattern of eclipses. Knowledge of the saros, as it applies to cycles of lunar eclipses goes back to very early times, probably preceding 1000 B.C.

satellite.   The astronomical name given to a smaller body revolving around a larger one (the primary body); for example, the Moon revolving around the Earth. The planets have a total of 36 known satellites with others suspected: Earth, one; Mars, two; Jupiter, fifteen; Saturn, ten; Uranus, five; Neptune, two, Pluto, one. Of these only Titan, of Saturn's family, possesses an extensive atmosphere which consists chiefly of methane.

satellite galaxy.   A small galaxy near a larger one and probably orbiting around it.

saturation.   (1) Applied to the atmosphere to mean the conditions when pressure of water vapor present represents equilibrium with water or ice surface. (2) In any phenomenon, the condition where increase in some important parameter yields no further observable change.

Saturn.   The second largest of the giant planets and the sixth in order of distance from the Sun. Saturn moves around the Sun at a mean distance of 9.5 A.U. once every 29.5 years, in an orbit with eccentricity 0.06 inclined 2.5° to that of the Earth. Its synodic period is 378 days. Its atmosphere is rich in methane (CH_4) and hydrogen (H_2). Saturn is yellowish in color and is among the brightest objects in the sky. See also Appendix.

Saturn Rings.   A beautiful system of small particles orbiting around Saturn giving the telescopic appearance of continuous rings. There are four main rings in the equatorial plane. The inner ring (crepe ring) is faintly visible only with a large telescope and is contiguous to the middle ring (bright ring). This in turn is separated from the outer ring by a distance of about 5000 km called Cassini's gap or division after its discoverer. The rings extend out to a distance of 135,000 km from the center of Saturn, and have a thickness of a few km. Saturn's rings are made up of a multitude of discrete particles revolving around the planet in a nearly circular orbit in the direction of Saturn's rotation. Radar and other studies indicate centimeter dimensions for most particles, but larger ones may be present in lesser numbers, ranging in size up to many kilometers. The rings are within Roche's limit, and may be remains of an unformed or shattered satellite. The particles are thought to be ice. (See Fig., p. 203.)

Saturn's satellites.   Saturn has 10 known satellites. The largest of these is Titan (2,850 miles in diameter), the only satellite in the solar system known to have clouds and a substantial atmosphere, probably methane. Eight of the moons are within 1 million miles of Saturn, one is a little more than 2 million miles, and Phoebe, the smallest, is over 8 million miles. Phoebe revolves from east to west as do Jupiter's outer moons. At

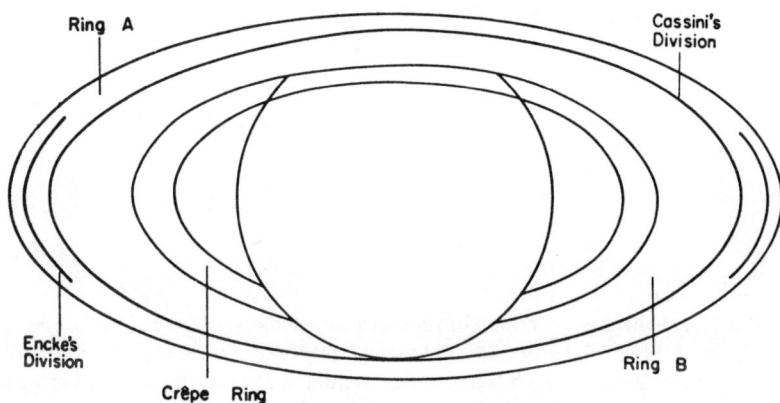

least one more suspected moon has been photographed inside Janus's orbit at the rings' outer edge.

saturnographic.   Referring to position on Saturn measured in latitude from Saturn's equator and in longitude from a reference meridian.

scalar.   A quantity which has magnitude but not direction. Time and mass are examples.

scale height.   The height at which a given parameter falls to a certain value (i.e., one scale height in density is when the density in the atmosphere falls to $1/2.72 (= 1/e)$ of its value at the surface).

scattering.   The absorption and reradiation of incident light by very small particles, or gas molecules. Scattering is greater for blue light than for red. The scattering particles must be smaller than the wavelength of the light which they are to scatter.

Schedar.   See Alpha Cassiopeiae.

Schmidt telescope.   A telescope or camera consisting of a spherical mirror with a "correcting plate" (thin lens) in front of it, generally with a small f-number and able to take a wide angle photograph. Invented by Barnard Schmidt, a German optician. (See Fig., p. 204.)

scintillation.   More usually known as "twinkling," the flickering of a star viewed with the naked eye, caused by heat currents in the atmosphere. Substantial scintillation, when the atmosphere is unsteady, is also known as poor "seeing" (q.v.). A similar related phenomenon is the scintillation of signals received from distant cosmic radio sources.

scirroco.   See sirocco.

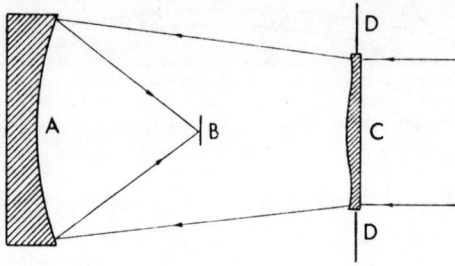

Scorpius (The Scorpion).   The eighth zodiacal constellation. The ecliptic passes only through the extreme Np corner. It is a striking group lying almost entirely in the Milky Way and one of the few which bears some resemblance to the object it is supposed to represent. The area is 497 sq degrees.

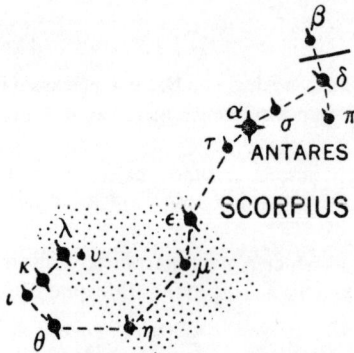

scud.   A popular name for the low drifting clouds which often appear beneath a cloud from which precipitation is actively falling. The official name for these clouds is fractostratus, or fractocumulus.

Sculptor (The Sculptor).   A drab constellation south of Cetus and Aquarius containing only four 4th magnitude stars. Formed by Lacaille in 1752 from a group of inconspicuous stars between Cetus and Phoenix, and has no distinguishing characters since the brightest stars are only of magnitude 4.5. The area is 475 sq deg. This constellation lies in a region rich in extragalactic nebulae, some of which are large and bright. Prominent among these is NGC 253, a beautiful example of an edgewise spiral. There is also the Sculptor galaxy, a widely dispersed and sparse aggregation of very faint stars which at 110,000 pc distance is one of the closest galaxies except for the two Magellanic Clouds.

Scutum (The Shield).   A constellation representing the shield of the Polish hero John Sabieski; it was introduced by Hevelius about 1660 and occupies a small area of 109 sq degrees in the Milky Way between Aquila and Sagittarius. (See Fig., p. 205.)

λ *Aquilæ*
M 11
(Wild Duck Cluster)

SCUTUM

M.17
(Omega Nebula)

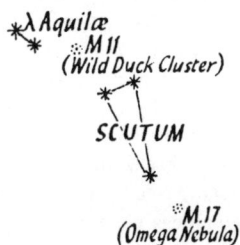

seas.   The dark areas or maria, of the Moon and Mars. The term dates from the 1600's when the dark areas were suspected to be oceans.

seasons.   The 4 divisions of any planet's year as defined by the passage of the planet through the 4 orbital points defining its 2 equinoxes and 2 solstices.

second.   A unit of time (1/3600 of an hour), abbreviated "sec". It equals the duration of 9,192,631,770 cycles of specified radiation from the cesium-133 atom. (2) A second of arc (1″) is a unit of angle (1/3600 of a degree). In 1 second of time the Earth rotates 15″ eastward and a star seems to move 15″ toward the west.

secondary cosmic rays.   Fragments of atoms created in the atmosphere stimulated by collisions between primary cosmic rays and atmospheric molecules or atoms.

secondary great circle.   A great circle perpendicular to a primary great circle.

secondary minimum.   The shallowest of the two minima in an eclipsing binary light curve.

secondary mirror.   The mirror, curved or plane, that reflects light into the eyepiece or on to other optical surfaces after it reflects from the primary mirror.

secular.   Nonperiodic, or constantly accumulating.

secular acceleration (of the Moon's mean motion).   A very slow shortening in the Moon's period of revolution round the Earth. It corresponds to a progressive increase of about 11″ a century in the Moon's mean motion.

secular parallax.   Change in the apparent direction of a nearby object due to the motion of the Sun through space.

seeing.   The term used to describe the quality of an image of a star or planet as affected by atmospheric conditions. The seeing is "bad" when air currents cause the image formed by the telescope to appear blurred, shimmering, or distorted; the seeing is "good" when the image appears sharp, steady, and clear.

seeing scale, Antoniadi.   The Greek astronomer E. M. Antonaidi, well known for his work in Mercury and Mars, produced the following seeing scale for planetary work: (1) Perfect seeing without a quiver. (2) Slight undulations, with moments of calm lasting several seconds. (3) Moderate seeing with large tremors. (4) Poor seeing, with constant troublesome undulations. (5) Very bad seeing, scarcely allowing the making of a rough sketch. The scale used in recording observations must be specified; the best way is to write, for example, Pickering's seeing (see next entry) 3 as 3/10, and Antoniadi as 3/5. Note that the two numerical sequences work in reverse order.

seeing scale, Pickering.   Scale devised by W. H. Pickering on the basis of observations carried out with a 5-inch refractor. Seeing 1-3 is considered very bad; 4-5 poor; 6-7 good; and 8-10 excellent. Pickering's scale, being based on the appearance of star disks, is not very suitable for planetary work where the observer is scrutinizing an extended image.

seleno.   Prefix from the Greek referring to the Moon.

selenocentric.   Relating to the center of the Moon or to the Moon as a center.

selenographic.   Of or pertaining to the physical geography of the Moon. Specifically referring to positions on the Moon measured in latitude from the Moon's equator and in longitude from a lunar reference meridian.

selenography.   The study of the physical features of the Moon's surface.

selenology.   The general study of the Moon, its magnitude, motion, composition, origin and the like.

semiregular variable.   A variable star in which the period is only approximately regular, often with a secondary period superimposed upon the primary cycle.

sensors, horizon.   A device for sensing the periphery of a body as seen from a satellite and then finding its geometric center. The direction from a satellite to the center of Earth or other body can be determined by means of this device.

separation.   The angular or linear distance between two objects.

Serpens (The Serpent).   Serpens is the serpent which Ophiuchus carries and appears on either side of him. The constellation is therefore in two parts, of total area 637 sq degrees. It is one of the star groups in the Almagest of A.D. 150 and Ptolemy assigned 18 stars to it. The first portion is a scattered star group north of Libra; the second is mainly in the Milky Way between Ophiuchus and Sagittarius where it is subject to heavy interstellar obscuration. There are 5 open clusters, but only one is an effective object— the remarkable NGC 6611 associated with the gaseous nebula IC 4703. All of the known planetary nebulae are small and difficult to see. (See Fig., p. 207.)

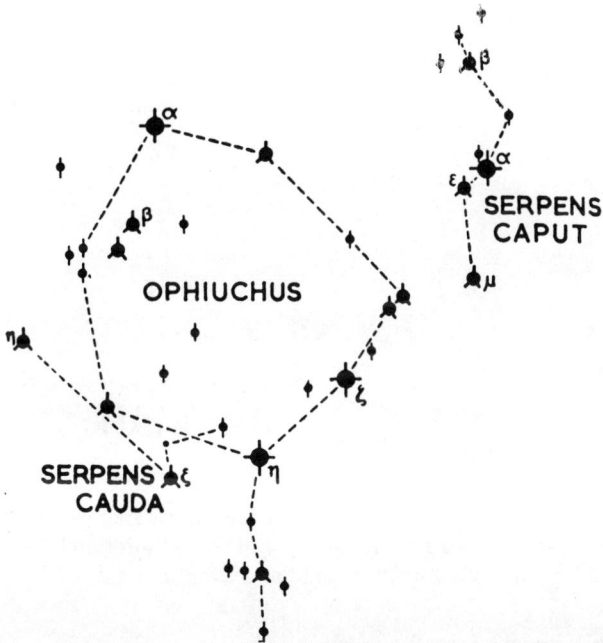

SERPENS
CAPUT

OPHIUCHUS

SERPENS
CAUDA

set.   Astronomers refer to the process of pointing a telescope on a star as "making a setting." A telescope is "set" when it is pointed on a star and tracking it automatically.

setting circle (telescope).   The circular scale used in a telescope to set it to the declination and right ascension of a celestial object.

Sextans (The Sextant).   Formed by Hevelius about 1680 in honor of the observatory instrument which he had used since 1658 (the nautical sextant was invented by Hadley in 1730). Sextans is a small constellation between Leo and Hydra and has no visual distinguishing feature. (See Fig., p. 208.)

sextant.   A double-reflecting instrument for measuring angles, primarily of altitudes of celestial bodies.

Seyfert galaxy.   Any of a class of galaxies marked by unusually bright, small nuclei, often bluish in color and emitting radio energy. They may be related to QSO's and quasars (q.v.) and may have explosive activity proceeding in their centers.

shadow.   Darkness in a region caused by an obstruction between the source of light and the region.

shadow transit.   The passage of a satellite shadow across the disk of Jupiter or Saturn.

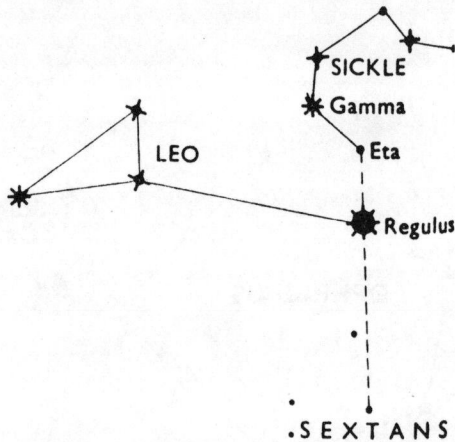

shell stars.   Stars, such as novae and supernovae, which are revealed spectroscopically as having rather broad emission lines, each with absorption lines and its violet edge. The lines are produced by an expanding shell of hot gas.

shooting star.   Popular term for meteor.

shock wave.   A sharp increase in pressure, density, and temperature created as a wave traveling through a gas or other medium and caused by an object or material moving through the medium faster than the speed of sound in the medium.

short-wave fadeout (SWF).   A short-wave fadeout effect caused by marked increase in the electron density in the D region and the lower E region. As a result, high-frequency radio waves which would normally pass through the D region and be reflected at high levels are absorbed. Long-distance radio communication that depends upon reflection in the ionosphere is therefore suddenly and completely disrupted for a period of about 15 minutes to an hour or so.

SI.   See International System of Units.

sidereal.   Referring to the stars, or measured with respect to the stars. Commonly applied to the rotation and revolution of the Earth, as in sidereal time, a 24-hour time system measuring Earth's rotation relative to the stars.

sidereal day.   The time interval between two successive passages of the vernal equinox across the local meridian. It equals approximately $23^h 56^m 4.09^s$ of solar time. Because of precession, this is 0.0084 sec less than the interval between two meridian passages of a fixed star.

sidereal month. The period of the Moon's revolution around the Earth with reference to the stars, 27.32166 days; the interval between two successive passages of the Moon through the same point in its orbit relative to the fixed stars.

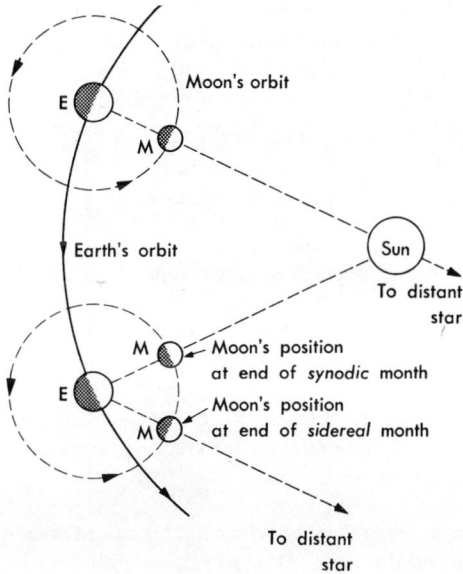

sidereal period. Rotation period of a body, measured with respect to the stars. The interval between two successive positions of a celestial body at the same point with reference to the fixed stars. The time taken by a planet or satellite to complete one revolution about its primary as seen from the primary and as referred to a fixed star.

sidereal rate. The rate at which the driving apparatus of a telescope must be set to keep a star centered in the eyepiece.

sidereal time. The hour angle of the vernal equinox, that is, the number of sidereal hours, minutes, and seconds that have elapsed since the vernal equinox was on the meridian. Alternatively, the sidereal time (S.T.) equals the right ascension (R.A.) of an object on the meridian. Sidereal time and mean solar time coincides only at the autumnal equinox, about September 21.

sidereal year. 365.25636 days. The time between two successive returns of the sun to the same position as marked by a line from the sun to a distant fixed star.

siderite. Old term for an iron aeleorite.

siderolite.   Old term for stony-iron meteorite.

sigma.   One standard deviation, q.v.

signal-to-noise ratio.   Ratio of strength of meaningful data from a source such as a star, to random noise arising within and outside the observing equipment.

sinking.   An atmospheric refraction phenomenon, the opposite of looming, in which an object on or slightly above the geographic horizon apparently sinks below it.

Sirius.   See Alpha Canis Majoris.

sirocco (sometimes scirroco).   A hot wind blowing in the warm sector of a cyclone. May be either dry or moist, depending on the type of air and its trajectory.

sleeks.   Minute, hairlike scratches on a telescope mirror arising from the use of a very hard pitch lap.

sleet.   Transparent globular hard grains of ice ranging in size from 1 to 4 mm (1/25 to 4/25 inch) which fall from clouds. They are usually produced as a result of the rain drops falling through subfreezing air.

slit.   A long, narrow opening through which light is passed in many spectroscopic instruments in order to produce optical images of spectral lines.

small circle.   The intersection of a sphere and a plane which does not pass through the center of the sphere, as a parallel of latitude. Any circle in a sphere smaller than a great circle. All parallels on a globe or map except the Equator.

Small Magellanic Cloud (SMC).   The smaller of two irregular galaxies believed to be satellites of the Milky Way. Visible from the southern hemisphere, the Small Cloud is at a distance of 200,000 light-years and about 10,000 light-years in diameter.

Sobieski.   Obsolete name of the constellation Scutum.

snow.   (1) White or translucent ice crystals or flakes of frozen water mainly in branched hexagonal shapes. (2) Loosely prepicitating crystals of other atmospheric gases, such as CO_2 on Mars.

snow stage.   The stage during the dynamic cooling of a mass of air during which solid water (snow) is sublimed into flakes.

solar.   Of or pertaining to the Sun or caused by the Sun (i.e., solar radiation, solar atmospheric tide). Relative to the Sun as a datum or reference (i.e., solar time).

solar activity.   Any type of variation in the appearance of the energy output of the Sun.

solar antapex. The direction toward which the nearby stars are apparently moving; the direction from which the Sun is moving relative to nearby stars. See following entry.

solar apex. The point on the celestial sphere toward which the Sun is traveling relative to nearby stars. The direction from which the nearby stars appear to be moving.

solar atmospheric tide. An atmospheric tide (pressure wave) due to the thermal or gravitational action of the Sun.

solar calendar. A calendar making the year conform as nearly as possible to the year of the seasons and neglecting the Moon's phases; its 12 months are generally longer than the lunar month.

solar cell. Cells that create electrical voltage from sunlight and which can be connected in large numbers to form a solar battery. If the surface of the solar cell is exposed to light, free electrons are produced in equal numbers as a consequence of the photoelectric effect by which solar photons knock electrons free. Solar cells usually have an exposed area of 1 or 2 square centimeters and the required number of cells are connected in series to supply the desired voltage.

solar constant. The amount of radiant energy received from the Sun in a unit of time upon a unit area of surface, at the mean distance of the Sun, at the top of Earth's atmosphere. Its average value is 1.39×10^{6} ergs/cm^{2}/sec. We speak of its "average" value since the solar "constant" may be slightly variable.

solar cosmic rays. Cosmic rays originating in the Sun.

solar cycle. The cycle defined by periodic increase and decrease in the number of sunspots and the reversal of their magnetic polarities. During a period of about 11 years, the number of spots waxes and wanes, with pairs of spots first having north magnetic poles on the leading side, and the latitude distribution changing. Then the polarity reverses with south magnetic spots leading for the next 11 years. Thus, the full solar cycle is about 22 years long, with two maxima and two minima in numbers of spots.

solar day. The interval between two successive transits of the Sun past a given meridian. The length of the solar day varies with the time of the year, but on the average it is 3 minutes 56 seconds longer than a sidereal day.

solar eclipse. See eclipse.

solar ecliptic limit. The solar ecliptic limit is the maximum angular distance of the Sun from the lunar node at which it can be just grazed by the Moon as seen from some point on the Earth. Within this distance the Sun will be eclipsed; but when the Sun is beyond this distance from the node, an eclipse cannot occur.

solar flares.   A region of exceptional brightness that develops very suddenly in a plage area of the chromosphere. Flares are often associated with sunspot areas of long life and with groups that have complex magnetic fields. There are several different types of solar flares, as indicated by their appearance and behavior. The brightness, dimensions, and duration are some factors taken into consideration in a system commonly used to classify flares according to their importance. Flares of minor importance are designated 1^-, whereas the largest flares are 3^+.

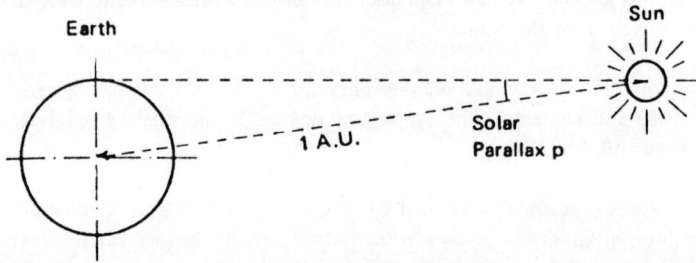

solar motion.   The motion of the Sun with respect to the nearby stars, about 19 km/sec.

solar parallax.   Defined as the angular equatorial radius of the Earth as seen from the Sun and measured as $8''.80184$. Below is a diagram of a solar parallax.

solar radiation.   Radiation emanating from the Sun, including electromagnetic, corpuscular, and infrared radiation.

solar rotation.   Circulation rate of solar surface features, varying with latitude as shown below. The rotation is fastest at the solar equator.

Solar Latitude	Rotational Actual	Period in Days Apparent (Earth viewpoint)
0	25.03	26.87
10°	25.19	27.06
20°	25.65	27.59
30°	26.39	28.45
40°	27.37	29.65
80°	33	—

solar system.   The system of the Sun and the bodies whose motions are determined primarily by the gravitational attraction of the Sun. The solar system is made up of thousands of particles and bodies varying in size from the Sun, which makes up about 99 percent of the mass of the solar system, down to the fine particles of dust floating in interplanetary space. Aside from the Sun the nine known planets are the other major bodies found in the solar system. The planets range from Mercury, the closest to the Sun at 36 million miles, to Pluto, the most distant at 3700 million miles from the Sun.

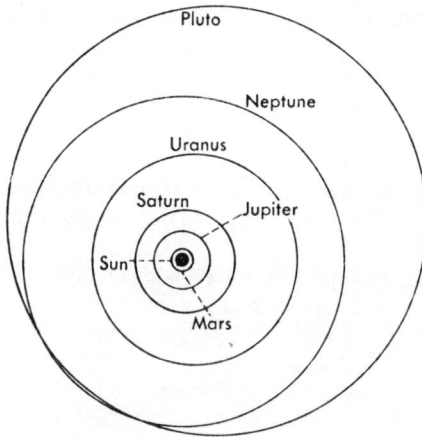

About two dozen planetary bodies exceed 1000 km in diameter. As shown below, it is difficult to show the orbits of the inner planets on a diagram showing complete orbits as large as Pluto's.

solar system origin.   The process of formation of the Sun and planets, dated at 4.6 billion (4.6×10^9) years ago, according to geochemical studies of meteorites and lunar samples. The following picture emerges from combination of astrophysical, geochemical, and planetary studies. As with other stars, the Sun formed from a contracting cloud of interstellar gas which probably broke up into a cluster of stars not unlike the Pleiades. The Sun was surrounded at this point by a disk-shaped cloud of leftover gas cooling from a temperature of a few thousand degrees. As this gas cooled, solid crystals condensed (just as snow flakes may condense from cooling air masses at high altitudes). At 1600° K, oxides of aluminum, titanium, and calcium condensed; at 1400° K, nickel-iron grains; at about 1300° K, silicate mineral grains; and at around 300 to 100° K, ice particles of frozen water, ammonia, and methane. These grains formed a dense layer in the central plane of the nebula, where gravity caused them to aggregate into objects of perhaps a few kilometers across. These planetesimals collided at relatively low speeds; smaller ones may have been fragmented but the larger ones tended to accumulate the smaller ones and their debris. Within a few million years (or perhaps much less) satellite- or planet-sized objects formed. The rest of the gas and dust was blown outward by force of the Sun's radiation pressure. Since dynamics forced the initial gas and dust to orbit in circular orbits, the consequent planets are in circular orbits. This theory, unlike some earlier theories that required rare accidents such as collisions of passing stars, accounts for planets and many of their properties as a natural outgrowth of formation of a single star (but not necessarily a binary star). See also nebular hypothesis.

solar time.   Time based upon the rotation of the Earth relative to the Sun.

solar wind.   Constant streams from the Sun of ionized particles (primarily hydrogen) which move outward with velocities of 400 to 800 km/sec (250 to 500 miles/sec) in the

Earth's vicinity. It contains about 5 particles/cm^3 and a magnetic field of roughly 10^{-5} gauss. Streams of these particles arise from sunspots and solar flares to cause magnetic storms and aurorae upon striking Earth. The solar wind of comparatively low-energy particles is a sporadic phenomenon rather than a steady breeze. By means of the solar wind, the Sun loses roughly 10^{-13} of its mass each year.

solar year.   The time between two successive returns of the Sun to the vernal equinox, or First Point of Aires. It is also called the "tropical year." It is 365.242 days.

solid angle.   The 2-dimensional or surface analog of angle. A measure of the number of square degrees or square radians in a part of the sky.

solid eyepiece.   An eyepiece in which the eye and field lenses are fused into a solid element.

solstice.   The dates each year when the Sun is farthest north (summer solstice) or farthest south (winter solstice) of the celestial equator. In the northern hemisphere the summer solstice (longest day) occurs about June 21, the winter solstice (shortest day) about December 22. Sunrise and sunset at these dates occur at extreme northerly or southerly points on the horizon; these points were utilized in calendric or other activities by ancient people, as at Stonehenge and in Peru.

solstitial colure.   That great circle of the celestial sphere through the celestial poles and the solstices.

south celestial pole (SCP).   The southern projection of Earth's axis upon the celestial sphere. No bright star like Polaris marks the location of the south celestial pole.

South Tropical Disturbance.   An elongated dark band on the cloud surface of Jupiter at about the latitude of the Great Red Spot. It was first seen in 1901 as a dark spot which then spread rapidly. It has at times exceeded 180° of longitude in length, and like the Red Spot it appears and disappears intermittently. It is probably a storm system in Jupiter's atmosphere.

southing.   A celestial object's crossing of the meridian.

Soyuz.   Russian word meaning "union," name of spacecraft used in docking experiments including Apollo/Soyuz mission to link with an American vehicle.

space.   The volume outside planetary and stellar atmospheres, extending in all direction with no known limits. The old term "outer space" may refer to space beyond the Earth's atmosphere, or space beyond the solar system. The term "deep space" often refers to space outside the solar system. Space is nearly a vacuum, with the gas density being of the order 10^{-23} gm/cm^3.

space coordinates.   A three dimensional system of Cartesian coordinates by which a point is located by three magnitudes indicating distance from three planes which intersect at a point.

space motion.   The velocity and direction of motion of a star with respect to the Sun. Also called space velocity. See also peculiar motion.

space polar coordinates.   A system of coordinates by which a point on the surface of a sphere is located in three dimensions: (a) its distance from a fixed point at the center, called the pole; (b) the colatitude or angle between the polar axis (a reference line through the pole) and the radius vector (a straight line connecting the pole and the point); and (c) the longitude or angle between a reference plane through the polar axis and a plane through the radius vector and polar axis.

space reddening.   The observed reddening, or absorption of shorter wavelengths, of the light from distant celestial bodies due to scattering by small particles in interstellar space.

spatial.   Pertaining to, characterized by, or occupying space.

specific gravity.   Density relative to the density of water, which is 1 gram/cm^3.

specific heat.   The amount of heat required to raise a substance's temperature 1°C relative to that required to raise the same amount of water 1°C.

specific humidity.   A measure of humidity. The mass of water contained in a given mass of moist air. Generally expressed in grams of water per gram or kilogram of moist air, sometimes called mixing ratio. Usually designated as q or Q.

spectral line.   A system of subdividing stars according to their spectra and thus essentially according to temperature. The patterns of absorption lines in stars' spectra are indicative of the various temperatures, sizes, densities and atmospheric conditions of the stars. A study of the spectra of many stars, both within and outside the Galaxy, has shown that they can be arranged in a definite sequence, generally known as the Draper (or Harvard) classification. Development of the system began in the 1880's when classes (A, B, C, etc.) were defined by appearance of hydrogen lines. Annie J. Cannon published a complete system based on 225,320 stars in 1918–1924, but further work rearranged the classes in order of temperature. The seven main groups or classes are indicated by the letters O, B, A, F, G, K, M. Each class can be subdivided to allow for finer gradations by adding a numeral from 0 to 9. Thus, B0 represents the first member of class B spectra and the last member is B9, followed by A0, and so on. A few stars have spectra that fall into the special categories R, N (or R-N), and S which form branches toward the end of the main classification. The following lists the characteristic features of spectral classes:

Class	Main Spectral Lines and Temperatures
O	Ionized helium, nitrogen, oxygen, and silicon; hydrogen weak. $\tau \simeq 40,000°K$
B	Neutral hydrogen and helium; ionized oxygen and silicon; ionized helium absent. $\tau \simeq 18,000°K$

Class	Main Spectral Lines and Temperatures
A	Hydrogen strong; ionized magnesium and silicon; ionized calcium, iron, and titanium begin to appear; helium absent. $\tau \simeq 10,000°$ K
F	Ionized calcium (Ca II) strong; some ionized and neutral metal atoms (iron manganese, chromium, etc.); hydrogen weak. $\tau \simeq 7000°$ K
G	Ionized calcium strong; neutral metal atoms increasing and ionized forms decreasing; molecular bands of CH and CN appear. $\tau \simeq 5500°$ K
K	Neutral metal atoms (including calcium) strong; molecular bands stronger; hydrogen very weak or absent. $\tau \simeq 4000°$ K
M	Neutral metal atoms very strong; TiO bands appear. $\tau \simeq 3000°$ K
R-N	Similar to K and M, but molecular bands of CH, CN, and C_2 strong; TiO absent.
S	Neutral metal atoms strong; oxide (ZrO, LaO, YO) bands strong.

spectral type.   The classification of a star according to certain spectral features.

spectrogram.   Photograph of a spectrum.

spectrograph.   A device for photographing spectra of light sources such as stars. The simplest of spectrographs consist of a prism placed in front of the lens of a camera designed to photograph stars. The prism refracts the starlight at different angles, blue light more than the red, therefore the point image of a star spreads out into a streak or spectrum from red to violet. Below is a schematic diagram of a spectrograph. In other spectrographs, a grating (q.v.) is used instead of the prism.

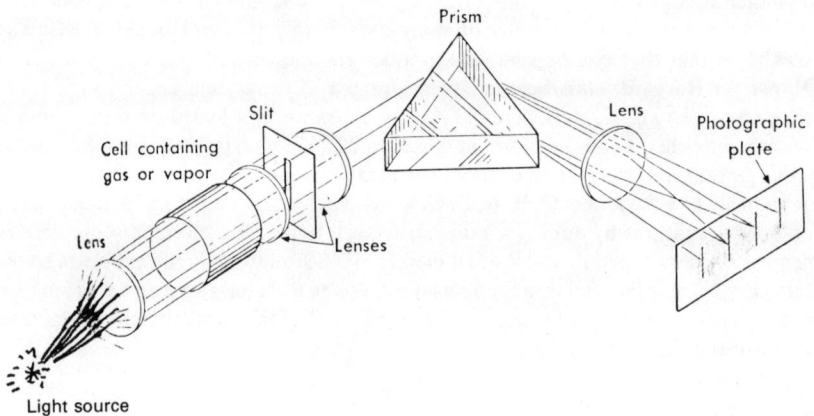

spectroheliogram.   A photograph made with a spectroheliograph (q.v.), showing the chromosphere and prominences taken outside eclipse in the light of a single spectral line (i.e., the light of a specific element in a specific state of excitation). These revealing photographs show how the gases of the element are distributed above the Sun's surface.

spectroheliograph. An instrument for photographing the Sun in light only from a specific spectral line. The most usual lines are the red lines of hydrogen Hx and the violet line of ionized calcium.

spectrohelioscope. An instrument for viewing the whole solar disk in light only from a specific spectral line. The instrument can detect the velocities of moving gases in the solar atmosphere.

spectrometer. An instrument used for display and precise measurement of spectra and their wavelengths. Instead of a photograph a spectrometer usually produces the spectrum as a graph plotting intensity versus wavelength.

spectrophotometer. An instrument for measuring and recording the amount of light (i.e., energy) at each wavelength in a spectrum, often in moderate-sized wavelength intervals (such as 100A) called "channels," or as a continuous plot of intensity vs. wavelength.

spectroscope. An instrument for observing spectra visually similar to a spectrograph, but with the photographic plate replaced by an eyepiece and the eye.

spectroscopic binary. A double star with components placed so close together that they cannot be separated optically, but having their association revealed by periodic Doppler line shifts in the spectral lines of their combined spectra caused by their orbital motion one around the other.

spectroscopic parallax. The parallax (or distance) of a star obtained by studying its spectrum. The spectral characteristics indicate roughly the luminosity of the star. Knowing this and also the apparent brightness of the star, its distance can be found by applying the inverse square law of illumination.

spectrum. The various colors of light from a source spread out in the sequence from red to violet (long wavelength to short wavelength), as in a rainbow. Invisible long wavelengths extend from the red to infrared and radio waves, and invisible short wavelengths from blue-violet to ultraviolet, x-rays, and gamma rays.

spectrum, absorption. See absorption line.

spectrum, continuous. See continuous spectrum, continuum.

spectrum, line. See lines.

spectrum, emission. See emission line.

spectrum variable. A star having a variable spectrum due to intrinsic varying properties of the star (such as changing magnetic field).

specular reflection. Reflection from a polished surface such as a mirror. Nearly all the light is reflected at the angle of incidence.

speculum metal. (1) An alloy of copper and tin used by early astronomers for telescope mirrors. (2) An early term for a telescope mirror, now obsolete.

speed. The numerical rate at which a body is moving without regard to direction of its motion. (Velocity includes rate of motion and direction of motion.)

speed of light. The speed of propagation of electromagnetic radiation. The speed in a perfect vacuum is a universal constant equal to 299,792.5 kilometers per second, or about 3×10^{10} cm/sec. The speed is slightly different in other media.

speed of sound. The speed of propagation of sound or any pressure wave in a medium. In dry air at 20°C it is 343 m/sec.

sphere. A three-dimensional figure produced by the revolution of a circle around its axis.

sphere of influence. The volume in space about a planet or star in which the planet or star is the dominant gravitational influence on a small body. For example, inside a planet's sphere of influence that planet dominates, while the Sun and other planets produce only relatively minor perturbative forces.

spherical aberration. The failure of rays reflected from a spherical surface to come to a common focal point. Spherical reflecting surfaces do not bring rays from different parts of the mirror to a focus in the same focal plane, which may also be a shortcoming of lenses. The effect of spherical aberration is a series of poorly defined images spaced along the axis of the objective. To produce a sharp, well-defined image, a mirror must be curved in the shape of a paraboloid, and lenses must have a combination of curves for the same result.

spherical angle. The angle between two intersecting great circles on a spherical surface.

spherical coordinates. A system of coordinates defining a point's position in terms of an imaginary sphere by giving the sphere's radius and the point's angular distances from a primary great circle and from a reference secondary great circle, as latitude and longitude.

spheroid. An ellipsoid; a figure resembling a sphere.

Spica. See Alpha Virginis.

spicules. Bright, short-lived jets rising above the lower chromosphere and seen especially on photographs of the solar limb. These hydrogen-emitting spikes are visible on the average up to 12,000 km above the photosphere; each lasts about five minutes and radial velocity measurements show up and down motions of 15 to 20 km/sec. The spicules are present in all parts of the limb at all times (at any moment there are at least

100,000 of them on the solar surface). They may be related to the short-lived granules of the photosphere, and they are probably an important factor in the transportation of energy upward to keep the corona heated.

spider.   The supporting structure for the diagonal or secondary mirror in a reflecting telescope.

spiral.   An abbreviated term for spiral galaxy (a galaxy having spiral arms).

spiral arms.   The concentrations of gas, dust, and bright, hot stars in the disk of a galaxy, arranged along spiral lanes leading out from the center. They are believed to be wave-like concentrations of gas, dust, and stars rotating with a period different from that of the galaxy itself (just as a wave may move across a river surface at a rate different from the river's motion).

spiral galaxy.   A galaxy of disk shape with spiral arms defined by stars, open clusters, and nebulae (as opposed to elliptical and irregular galaxies).

spiral nebula.   Obsolete term for a galaxy with a spiral structure, dating from before the 1920's when galaxies were thought to be gaseous nebulae inside our own Milky Way.

sputnik   Russian term meaning "companion," given to the first artificial Earth satellites.

squall.   A sudden, violent rain or snow storm accompanied by strong, gusting winds.

stability.   The state in which a body or system tends to return to its initial configuration after a small disturbance.

standard atmosphere.   A hypothetical state of the atmosphere in which the pressure, temperature, lapse rate, etc., are given to serve as a standard of comparison in various meteorological calculations.

standard error (s.e., sometimes called standard deviation).   A statistical measurement of the accuracy with which a quantity is known. It is often designated σ, or "sigma." In the normal case, there is a probability of about 68 percent that the true value is within one s.e. of the measured value. The standard error is about 1.48 times the probable error (q.v.).

standard time.   The local mean solar time at a standard meridian, adopted over a zone for civil convenience.

star.   A self-luminous body that generates energy by nuclear reactions within itself, which it radiates continuously in all directions into its surroundings. The individual character of a star depends on its composition and mass, which determine other properties such as size, surface temperature, and luminosity.

star catalogue.   A listing of stars giving observational data, such as positions for a specified mean equinox and equator.

star clouds.   Portions of the Milky Way, so called because the stars are so numerous and close together that they look like luminous clouds. The brightest clouds are located in the constellations Sagittarius and Scutum in the densest regions of the Milky Way. The term is dying out, since the clouds have been identified as heavily populated but ordinary portions of spiral arms.

star clusters.   A group of stars physically close together and gravitationally associated. There are several recognized types of clusters which range in numbers of stars from loose aggregates of a few stars to vast collections of stars too numerous to count. See globular clusters and open clusters.

star clusters, globular.   See globular clusters.

star clusters, open.   See open clusters.

star nomenclature.   The brightest stars are given Greek letter prefixes in approximate order of their brightness, followed by the name of the constellation. After Greek letters small English letters and finally English capitals are used. Also, stars down to the 6th magnitude have numbers, called Flamstead numbers, within each constellation. All stars are also named by the declinations and a number in the Bonn Durchmusterung. Other catalogs, often of special star types, provide additional designations.

starquakes.   Hypothetical crustal fractures or movements of interior mass causing observed small but sudden changes in rotation rates of neutron stars.

star streams.   Systems of relatively nearby stars with preferential directions of apparent motions associated with differential rotation of the galaxy. They are revealed by statistical studies of stars' peculiar velocities.

star trackers.   Sensors that detect the direction of a given star (or group of stars) and provide a supervisory function for a gyro-stabilized inertial platform. They are often based on frequency-modulated disk-scanning methods.

stars, brightest.   See Appendix.

stationary front.   Weather system fronts that have little movement. A front of this type may develop either into a cold or a warm front but more often resembles a warm front.

static.   (1, noun). Radio signals interfering with the signal that is desired. (2, adj.) Pertaining to a model or theory which ignores evolution and analyzes only the structure of phenomenon of a fixed point in time.

steady-state theory.   A model for the universe developed by H. Bondi, T. Gold, F. Hoyle, and others, especially in 1950 and 1960's. Assuming the perfect cosmological principle that the universe appears approximately the same in all times and places, they assume that the universe has no end and no beginning and retains a constant mean density. Since the redshifts of galaxies show the universe to be expanding, they require that matter be created at a very small rate (about 3×10^{-45} g/cm^3/sec) to keep the density constant. Discovery of the 3 K radiation predicted by the big-bang theory and not predicted by the steady-state theory has caused most cosmologists to abandon the steady-state model.

Stefan's law.   The total energy radiated in each second from a body of temperature T and area A is $E = \sigma T^4 A$, where σ is the Stefan-Boltzmann constant, 5.67×10^{-5} ergs/cm^{2}K^4 sec.

stellar associations.   See associations.

stellar atmosphere.   The outer, tenuous gas of a star, usually the region above the photosphere (opaque layer), but sometimes intended to mean all outer regions of stars, especially in cases where (unlike the solar case) the photosphere is poorly defined. Described in terms of pressure, temperature, density, and opacity at different heights.

stellar energy.   The temperature of stellar interiors (including the Sun) can be estimated from the physical conditions at the surface and from a knowledge of the laws of physics, to be of order of 15×10^6 degrees centigrade for most stars (on the main sequence). There are two hydrogen-fusing reactions that provide this energy, the proton-proton chain (q.v.) and carbon cycle (q.v.). Both reactions transmute four hydrogen nuclei into one helium nucleus, and when this occurs, about a half of one per cent of the mass of hydrogen disappears and is converted into energy. More evolved stars, such as giants, create their energy by high-temperature reactions that fuse heavier elements, such as helium.

stellar evolution.   See evolution, stellar.

stellar interferometer.   An instrument which uses the interference of light to measure very small angles between luminous objects (i.e., separation of binary stars of the diameter of stars).

stellar interior.   The inner part of a star or a model thereof, usually described in terms of temperature T, pressure P, density ρ, composition, and energy generation rate at various points from the center to the surface.

stellar magnitude and brightness.   See magnitude.

stellar model.   A tabulation of temperature, density, pressure, and related properties at various levels inside a star and its atmosphere, based on theoretical calculations.

stellar parallax.   The angle subtended by one a.u. at the distance of the star being studied.

stellar spectra.   See spectrum, spectral classes.

stereographic projection.   Map of a sphere as projected upon a target plane from a point diametrically opposite to the point of tangency.

stony meteorite.   A meteorite composed of silicate rock material without major regions of nickel-iron metal, which are found in "stony-iron" and "iron" types of meteorites.

stooping.   A type of mirage; a special case of sinking in which the curvature of light rays due to atmospheric refraction decreases with elevation so that the visual image of a distant object is foreshortened in the vertical.

Straight Wall.   A prominent linear cliff on the Moon, believed to be a fault scarp.

stratocumulus.   A cloud layer, or patches of clouds, composed of laminae, globular masses, or rolls; the smallest of the regularly arranged elements are fairly large; they are soft and gray with darker parts. These elements are arranged in groups, in lines, or in waves, aligned in one or in two directions. Very often the rolls are so close that their edges join; when they cover the whole sky they have a wavy appearance.

stratopause.   The upper limit of the stratosphere and the lower limit of the mesophere in the Earth's atmosphere.

stratosphere.   (1) An upper portion of a planetary atmosphere, above the troposphere and below the ionosphere, characterized by relatively uniform temperature and horizontal winds. On earth, its lower limit varies from about 8 to about 20 km; its upper limit lies at about 25 kilometers. The temperature in this region is about $-75°C$. The base of the stratosphere makes an upper limit to the general turbulence and convective activity of the troposphere; thus air motion within the stratosphere is largely horizontal (jet stream). (2) In some usages, the region above the troposphere all the way up to a point where a strong temperature increase begins, at about 100 km altitude.

stratus (clouds).   A low uniform layer of cloud resembling fog, but not resting on the ground. Another word for layer, perfectly describing the clouds which appear in "straight" sheetlike cloud layers.

Strömgren sphere.   Theoretically, a spherical region of ionized hydrogen (an H II region) around a bright hot star. In practice the regions may be more ragged than spherical due to dust, gas motions and nonuniform density. The diameter may reach 325 light-years for an 06 star, 110 for a B0 star, and only 10 for a B9 star, due to the rapid falloff in the ionizing ultraviolet radiation from cooler stars.

subgiant.   A star lying between the normal giants and the main sequence in the H-R diagram.

sublunar point.   The point on the Earth at which the Moon is in the zenith at a specified time.

subsolar point.   The point on the Earth at which the Sun is in the zenith at a specified time.

substellar point.   The point on the Earth at which a star is in the zenith at a specified time.

subtend.   To extend in angular measure, as in "the moon subtends an angle of 1/2 degree."

sudden ionosphere disturbance (SID).   An ionospheric disturbance which manifests itself within a few minutes of the appearance of some strong solar flares as a sharp fade-out of long-distance, short-wave (i.e., high frequency) radio communication on the sunlit side of the Earth.

summer solstice.   The date at which the Sun is farthest from the equator and in the hemisphere of the observer. It occurs about June 21 in the northern hemisphere and December 22 in the southern hemisphere. The longest day of the year occurs on this date.

Sun.   The central star that dominates the entire solar system, its diameter is 109 times that of the Earth and its volume, 1,300,000 times. Its mass is 332,000 times that of our planet, or about 700 times the combined mass of all the planets, or 1.99×10^{33} grams. The central temperature is about 15 to 20 million K; the surface temperature is about 6000° K. Its surface gravity is 27.9g, and its period of rotation, about 27 days.

sun pillar.   A vertical shaft of white light extending above and below the Sun, occasionally seen during very cold weather. It is caused by reflection from tubular shaped ice crystals in the atmosphere.

sundial.   An instrument for showing the time by means of a shadow cast by the Sun on a dial plate.

sunrise.   The crossing of the visible horizon by the upper limb of the ascending Sun.

sunset.   The crossing of the visible horizon by the upper limb of the descending Sun.

sunspots.   Temporary dark regions on the Sun, consisting of an inner, quite dark, core (umbra) and a somewhat lighter surrounding area (penumbra). They appear dark only because they have a lower temperature (4000 to 7000 K) than the surrounding photosphere. Often appearing in pairs with opposite magnetic polarity, they are magnetic disturbances in the photospheric gas. Some have magnetic fields of the order of 3000 to 4000 gauss at the center of the spot. The giant sunspot of April 1947 had a maximum measured area of 6132×10^{-6} of the Sun's hemisphere. The number of sunspots varies from day to day and from one year to the next. The spots come and go in marked cycles, with pronounced maximum numbers of around 100 spots on the visible disk occurring at intervals of about 11 years. Magnetic polarities also reverse every 11 years, so that a complete cycle requires 22 years.

sunspot cycle.   See sunspots and solar cycle.

supercluster.   A cluster of clusters; usually with reference to galaxies.

superdense stars.   Stars of extreme density, with normal stellar mass but the size of small planets or asteroids. Superdense stars are the result of supernovae. When a star runs out of nuclear fuel, no more heat is generated to keep the gas expanded. A complete collapse of the stellar core accompanies the violent explosion of the outer layers, which go to form a nebular shell. The white dwarfs represent one class of dense stars, in which electrons interact and provide the pressure to support the outer layers. White dwarfs have earthlike dimensions. Still more massive stars have more gravity and collapse down to asteroidal dimensions whereupon protons and electrons are jammed together to form neutrons. The star is thus composed largely of neutrons and is called a neutron star. Many of these are pulsars. Superdense stars generate little or no heat of their own, but remain hot for a long time as they radiate their original heat.

supergiant.   A star so great in size that it could contain a large part of the solar system, and so luminous as to equal some 100 to 10,000 suns. They are generally evolving very rapidly, on a time scale of centuries or decades.

supergranulation.   Large scale granulation of the solar surface caused by convective cells up to 30,000 km across.

superior conjunction.   The position of a planet when it lies on the far side of the Sun from the Earth, so that the three bodies are in a line when seen from "above" the ecliptic plane.

superior planets.   Planets whose orbits lie outside the Earth's orbit. Below are the aspects and phases of a superior planet. The aspects are similar to those of the Moon. The only phases are full and gibbous.

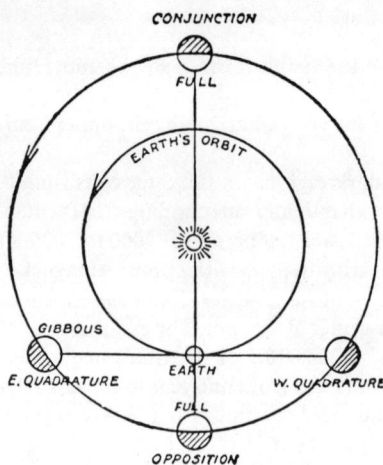

supernova. A massive star that explodes catastrophically with a sudden increase (of a billion times) in brightness and a liberation of its energy. The explosion is due to instabilities inside the star as it runs out of nuclear fuel. It is about ten thousand times as bright as an ordinary nova with an absolute magnitude of -16 or -17, releasing as much energy in one second as the Sun does in 10 to 100 years. The remains of the supernova of A.D. 1054 are still visible as the irregular, expanding Crab Nebula. The supernova of 1702 is Cassiopeia A, the brightest radio star. Only four supernovae have been seen in our own galaxy during the past thousand years. These were the stars of 1006 (in Lupus); 1054 (in Taurus); the Crab Nebula supernova; in 1572 (Tycho's Star in Cassiopeia); and 1604 (Kepler's Star in Ophiuchus). None has been seen since the invention of the telescope, but many have been seen in other galaxies. They have been divided into subtypes with different rates of brightness change.

surface gravity. The gravitational acceleration exerted by a planet or a body on a small particle at its surface. The gravitational acceleration depends upon the mass (M) of the planet and its radius (R), and is given relative to Earth's surface gravity, by M/R^2 where M and R are in terms of the mass and radius of Earth.

Swan bands. Spectral bands created by the carbon radical, C_2, common in several celestial objects such as carbon stars. They were first studied by W. Swan in 1856.

symbiotic stars. Binary pairs of stars that seem to occur together and may involve mass transfer from a larger star to a small, hot, blue variable which may be in a prenova state.

synchrotron radiation. Radiation from relativistic electrons (velocities approaching the velocity of light) spiraling in a magnetic field. The radiation is a continuum in either the optical or the radio regions of the spectrum, depending on the electrons' energies.

synchronous. Referring to the condition where the rotation of one body in a coorbiting pair takes the same time as its orbital revolution around the other body, so that one body keeps the same face toward the other. The Moon, for example, is synchronous with earth.

synchronous orbit. An orbit with the revolution period equal to the rotation period of the primary, so that the orbiting body stays above the same longitude on the primary.

synodic month. The interval between successive conjunctions of the Moon and Sun from new moon to new moon again. This month of the phases is longer than the sidereal month by more than two days. The length of the synodic month averages $29^d12^h44^m2^s$ (or a little more than 29 1/2 days) and varies more than half a day during the year.

synodic period. The period which elapses between two successive conjunctions or oppositions of a planet with the Sun as seen from the Earth. It is the interval after which the faster-moving inferior planet again overtakes the Earth, or the Earth again overtakes a slower outer planet. Mars and Venus have the longest synodic periods of the principal planets. The synodic periods of the outer planets approach the length of the year, since their orbital motions are very slow.

synoptic. Made from data compiled over a wide region at essentially the same time (i.e., a weather chart showing weather systems at a given moment).

Syrtis Major. Generally considered the most prominent and easily recognized of the dark markings on Mars, and probably the marking shown on the earliest drawings made in the 1600's. It has roughly a triangular shape with one tip pointing north.

system of astronomical constants. An interrelated group of values constituting a model of the Earth and the motions which together with the theory of celestial mechanics serves for the calculation of ephemerides.

Systems I and II. On Jupiter (and sometimes Saturn), systems of longitude for defining positions of equatorial and higher-latitude cloud systems (respectively), which circulate at different rates.

syzygy. (1) A point of the orbit of a planet or satellite at which it is in conjunction or opposition. When the Earth, Moon, and Sun lie on the same line; more correctly stated when the centers of the Earth, Moon, and Sun lie in a plane vertical to the plane of the Earth's orbit. (2) An alignment of three or more cosmic bodies in space.

T

T association. An association (q.v.) rich in T Tauri stars.

3C catalogue. The third Cambridge catalogue of radio sources.

30 Doradus. Part of the great looped nebula (Tarantula Nebula) in the Large Magellanic Cloud (LMC). The nebula is much more extensive and luminous than the Orion Nebula, but the Cloud of Magellan is about 113 times further away, making 30 Doradus appear faint. It has been interpreted as a partially formed nucleus of the LMC galaxy.

T Corona Borealis. One of the most erratic variable stars, aptly named "the Blue Star." Normally of the 9th magnitude, in May 1866 it shot up more or less overnight to the 2nd magnitude. After a week of naked eye visibility it vanished from view once more, returning to its original brightness until February 1946 when it surged up to the 3rd magnitude. Once more it faded rapidly and is now viewed as a magnitude 9.5 object about a degree south of Epsilon.

T Orionis variable stars. These stars have light curves similar to those of T Corona Borealis, marked by irregular fadings normally of two or three magnitudes. They have quite high surface temperatures between 11,000° and 25,000°C. They also seem to be giant stars rather than dwarfs. From the evidence available it appears quite probable that these are young stars condensing out of the vast gas clouds in which they are bedded.

T Tauri variable stars. Stars which exhibit erratic variations and have luminosities comparable to that of the Sun. They tend to occur in groups and are always associated with regions full of interstellar nebulosity, dust, and ongoing star formation. Their spectra are peculiar with bright lines of hydrogen, indicating that mass is being blown away from these stars. They are believed to be newly formed stars just blowing away the surrounding debris of their own formation, and are one of the most widely discussed types of variables. Many have surrounding clouds of dust that may be sites of planet formation. Sometimes called RW Aurigae stars.

tangential velocity. The component of velocity of a body at right angles to the line of sight, measured relative to the Sun in km/sec.

Tarantula Nebula. See 30 Doradus, above.

Tau Ceti. A star of magnitude 3.5 in the southern part of Cetus the Whale. It is one of the nearest stars being only 12 light-years distant. Tau Ceti has spectral type G8, nearly the same type as the Sun. The question has thus been raised whether it might have a family of planets resembling our solar system; it is therefore one of the stars selected for observation by radio telescope in an attempt to detect interstellar life. Early results have been negative.

Taurus (The Bull).   The second sign of the Zodiac and the third constellation. For many years Taurus was honored and worshipped as the leader of all the heavenly hosts. The constellation is rich in myths and legends from many lands and ancient sources. It is especially associated with the Mediterranean bull cult. It is a large, conspicuous group introduced by the Pleiadis rising in the east, followed by the triangular Hyades containing the brilliant Aldebaran. The area is 797 sq degrees and there are nine open clusters including the well-known moving star groups of the Pleiades (414 light-years distant) and the Hyades (137 light-years distant).

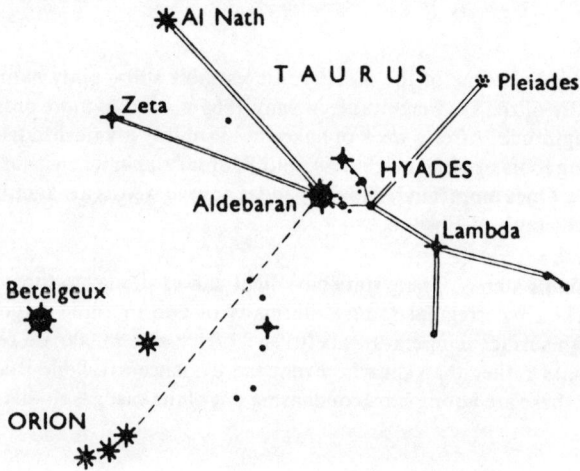

tektites.   Glassy material (Greek: *tektos*, "molten"), probably resulting from meteorite impacts but thought by a few researchers to be of extraterrestrial (lunar?) origin. Tektites are commonly about 2 or 3 centimeters across, although some are much smaller or larger. They are often rounded and often spheroidal in form; their glassy bubble nature and general appearance strongly suggest that they were formed by the melting of mineral matter. Tektites are largely restricted to strewn fields in a few regions on Earth. It is now generally believed that they were blown into these regions in jet-like swarms blasted away from terrestrial sites where large meteorites struck and fused local material. Some tektite fields have the same age as nearby large, ancient impact craters, from which they probably derived.

telemeter.   To transmit instrument recordings from a distance by radio.

telescope.   A telescope consists chiefly of two optical parts; an objective, or large light-gathering and focusing device, such as a lens or mirror, and an eyepiece, or small focusing device, usually a lens. The two are held together by a tube or framework that can be pointed in the desired direction. Since all astronomical objects lie at great distances, the light rays from them are practically parallel as they fall upon the objective of the telescope. The objective then brings them to a focus in the image located in the focal plane. The eyepiece is placed so that its distance behind the focal plane is equal to its own focal length. Thus, the diverging light rays from the focal plane pass through the eyepiece and again become parallel as they enter the observer's eye. See also reflector, refractor, eyepiece.

telescope, radio. See radio telescope, radio astronomy.

telescope, reflecting. See reflector.

telescope, Schmidt. See Schmidt telescope.

telescope, X-ray. See x-ray telescope.

Telescopium (The Telescope). A small, uninteresting constellation near Pavo, with no star brighter than magnitude 3.8. It was formed by Lacaille in 1752 between Pavo and Sagittarius. The area is 252 sq degrees and there are a few double stars, a single globular cluster NGC 6584, and one planetary nebula.

telluric. Terrestrial in origin. Especially, spectral bands formed by gases in our atmosphere rather than in an astronomical object being observed.

temperature. A measure of the internal energy per particle of a body (i.e., the kinetic energy due to microscopic motions of constituent particles). Astronomers use many different measures to determine temperature, distinguished as color temperature, brightness temperature, ionization temperature, etc. All of these are measures of the mean kinetic energy (due to velocities) of gas atoms and ions. In any gas, an average particle has an energy given by $3/2 \, kT$, where k is the Boltzmann constant, 1.38×10^{-16} erg/degree. Temperature is usually measured in science on a scale of degrees Kelvin (abbreviated K) where 0 K is an idealized state with zero internal energy. Water's freezing point is 273 K; boiling is 373 K. The centigrade or Celsius scale (°C) is Kelvin $-$ 273. Fahrenheit is $9/5°C + 32$. There is no theoretical upper limit to temperature.

template. A curve of known radius used to measure the depth or curve of a mirror.

terminator. The line which divides the dark from the illuminated part of the disk of a planet or of the Moon. The line of sunset or sunrise on the Earth, Moon, or planets.

terrestrial planets. The planets of the solar system of about the same size and silicate composition mass as Earth; Mercury, Venus, Earth, and Mars. In old lists, Pluto was sometimes included because of its small size, but it is now thought to be fundamentally different from the terrestrial planets listed above, and is no longer included.

terrestrial radiation.   The total infrared radiation emitted from the Earth's surface; to be carefully distinguished from effective terrestrial radiation, atmospheric radiation (which is sometimes used erroneously as a synonym for terrestrial radiation), and insolation.

terrestrial refraction.   See refraction of the atmosphere.

terrestrial space.   **(1)** The region near the Earth. **(2)** The region between about 200 miles altitude and a few Earth radii above the Earth's surface.

Tethys.   An inner satellite of Saturn, with a diameter of about 1000 km and an orbital radius of 295,000 km.

theodolite.   An instrument consisting of a telescope mounted so that it can be revolved about both its vertical and horizontal axes for measuring the vertical and horizontal angles to various points. Used for surveying and for tracking moving objects such as rockets and balloons.

thermal diffusivity.   The process of heat distribution and diffusion.

thermal expansion (telescope).   The expansion of a mirror with an increase in temperature.

thermal noise.   Minor electrical signals generated in an electronic instrument by thermal motions of atoms and electrons. It limits the sensitivity of the instrument and can be reduced by cooling the instrument with dry ice or liquid nitrogen.

thermal radiation.   Radiation received from a body as a result of its own heat, in accord with the radiation laws, but excluding other types of radiation such as synchrotron radiation. By Wien's law, thermal radiation from planets and circumstellar dust is generally infrared radiation. Any body radiates; the higher the temperature, the bluer the radiation. Cool stars are thus red; hot stars, blue. See also black body, Stefan's law.

thermodynamics.   A branch of physics dealing with the study of heat and heat transfer.

thermogram.   The record of temperature as recorded by a thermograph.

thermograph.   An instrument designed to record the temperature over a period of time.

thermosphere.   An atmospheric layer at altitudes above 80 km on Earth, where temperature increases up to a height of roughly 320 to 400 kilometers. The temperature attained in the thermosphere is highly variable because it changes with time of day and, to some extent, with the latitude; it is also dependent on solar activity.

thindown.   The process by which cosmic rays, meteoroids, etc. lose their identity or their force as they penetrate into the Earth's atmosphere. The expenditure of heavy primary cosmic-ray energy in ionizing the substance (usually air) through which it passes.

Third Quarter.   The phase of the Moon when it has completed three-quarters of one complete revolution round the Earth from the Earth–Sun line. It then appears as a half moon in the morning sky.

three-degree background radiation.   Microwave radiation seen uniformly in all directions. Discovered in 1964, it matches red-shifted radiation predicted to result from the big bang and strongly supports the big-bang theory.

three-kiloparsec arm.   A hydrogen cloud located about 3 pc from the galactic center and expanding outward toward the solar system (located 9 to 10 kpc from the center). It is believed to be part of an inner spiral arm.

tidal hypothesis.   A theory involving the close approach of two bodies with the partial tidal disruption of at least one producing planets, satellites, or rings. The theory has been rejected as applied to origin of the solar system, but may be relevant to ring systems or some satellites.

time.   The quantity measuring duration of events. It is commonly reckoned by the position of a celestial reference point relative to a reference celestial meridian. Time may be designated solar, lunar, or sidereal as the reference is the Sun, Moon, or vernal equinox, respectively. Solar time may be further classified as mean or astronomical if the mean sun is the reference, or as apparent if the apparent sun is the reference. See mean solar time, sidereal time.

time zone.   A longitudinal division on the Earth's surface roughly 15° wide in which civil time is taken as a whole number of hours earlier or later than G.M.T.

Titan.   The sixth moon of Saturn, and the largest satellite in the solar system. It is believed to be covered by reddish clouds and may be the only satellite with a massive, cloudy atmosphere. Its diameter is about 4850 km, similar to Mercury's.

Titania.   A satellite of Uranus orbiting at a mean distance of 438,000 km. Its diameter is roughly 2000 km.

Titius.   Bode rule. See Bode's rule.

tool.   The glass disk with which a telescope mirror is ground.

Toro.   Asteroid 1685, which crosses Earth's orbit and can approach to within 0.13 A.U. Spectral and radar observations suggest a thin dust layer over a rocky surface, which is one of the closest known spectral matches to certain types of chondritic meteorites. These meteorites may thus actually be pieces knocked off Toro onto Earth-impacting orbits. It is roughly 5 km across.

toroid.   A doughnut-shaped volume generated by the rotation of a plane closed curve about an exterior axis lying in the plane.

torr.   A unit of atmospheric pressure equaling 1 mm of mercury or 0.0132 atmospheres.

total eclipse.   An eclipse during which the light of one body is completely cut off.

towering.   A refraction phenomenon; a special case of looming in which the downward curvature of the light rays due to atmospheric refraction increases with elevation so that the visual image of a distant object appears to be stretched in the vertical direction. The opposite of towering is stooping.

train.   Anything (such as luminous gas or ionized particles) left along the trajectory of a meteor after the head of the meteor has passed.

transit.   (1) The passage of a small body across a large body, as with Mercury transiting the Sun. (2) The moment when a star or planet crosses the meridian. (3) The moment when a detail on a planet's disk is carried across the planet's meridian by the planet's rotation. (4) The passage of a satellite's shadow across the disk of its primary planet; especially in the Jupiter system.

translunar space.   Space lying beyond the orbit of the Moon.

transmission grating.   A diffraction grating designed to transmit light instead of reflecting it.

transverse projection.   A map projection which is turned 90° from its usual orientation, and consequently is centered upon some other great circle than a meridian.

transverse velocity.   Motion at right angles to the line of sight usually given in kilometers per second.

transverse waves.   Waves in which the direction of vibration is perpendicular to the direction of transmission of energy as in the case of light waves and sound waves in a solid (not in a gas or liquid).

trapezium.   (1) A four-sided figure having no two or only two parallel sides. (2) The group of four stars at the heart of the Orion nebula, or similar groups in other nebulae.

triangulation.   Determining the distance to an inaccessible object by measuring its direction from the two ends of a base line of known length. Below triangulation is used to find the distance to a nearby star. To obtain the parallactic angle π, measurements are made when the Earth is at A and six months later when it is at B. See also parallax. (See Fig., p. 233.)

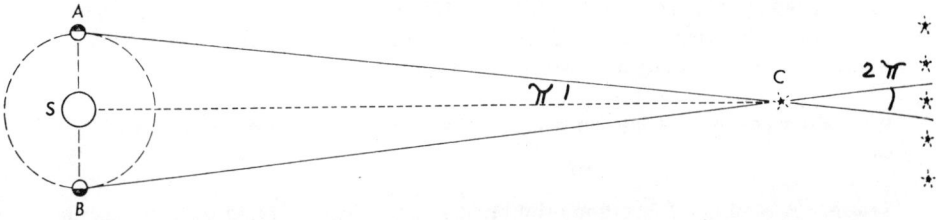

Triangulum (The Triangle).   The group of stars known as the Triangulum may be found between Aries and Andromeda. The Romans called the group Deltotum, the Delta; and it was associated with Egypt and the delta of the Nile. The three principal stars form a well defined triangle, easily seen and interestingly placed among the larger constellations. Alpha and Beta were widely known as the "Scale Beam." It has an area of 132 sq degrees. The chief interest in this constellation centers on NGC 598, a large spiral galaxy. It is a member of the local group of galaxies including the Milky Way, and in it may be well studied the component stars, clusters, and diffuse nebulae characteristic of these objects.

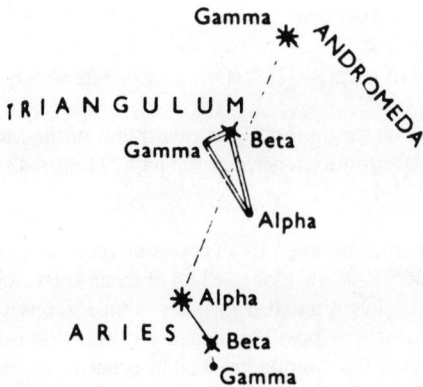

Triangulum Australe (The Southern Triangle).   A distinctive constellation which contains no bright stars. Alpha has magnitude 1.9. The other two members of the triangle are of the 3rd magnitude. Triangulum Australe lies close to α Centauri. One open cluster, NGC 6025, is distinctly visible with the naked eye on a clear night. The area is 110 sq degrees.

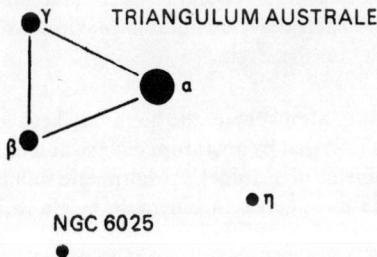

Trifid nebula.   A large emission nebula (M20, NGC 6514) about 3300 light-years away. It has a diameter of about 13 light-years and a mass of about 150 solar masses. Many newly formed stars are associated with it.

triple-alpha process.   A nuclear reaction, in which helium is consumed and fused into carbon.

Triton.   A satellite of Neptune orbiting at a mean distance of 354,000 km. Its diameter is large but uncertain (perhaps 4000 to 6000 km), making it one of the largest, if not the largest, satellite.

Trojans.   Two groups of asteroids, one 60° in front and the other 60° behind Jupiter, as seen from the Sun moving in the same orbit. The largest Trojan asteroid, Hektor, is about 100×300 km in diameter. Many smaller ones are known. They are composed of dark-colored material (albedo only 0.02 to 0.03), and their origin is uncertain.

tropic.   The region within $23\,1/2°$ of the equator in which the Sun can reach the zenith on some day of the year. The northern boundary is called the Tropic of Cancer, and the southern, the Tropic of Capricorn.

tropical.   (1) Relating to tropics. (2) Relating to or referenced.

tropical month.   The average period of the revolution of the Moon about the Earth with respect to the vernal equinox, a period of 27 days, 7 hours, 43 minutes, 4.7 seconds or approximately $27\,1/3$ days.

tropical year.   The interval between two successive returns of the Sun to the vernal equinox. Its length is $365^d5^h48^m46^s$ ($365^d.24220$) of mean solar time and is now diminishing at the rate of $0^s.53$ a century. It is the year of the seasons, the year to which the calendar conforms as nearly as possible. Because of the westward precession of the equinox, the Sun returns to the equinox before it has gone completely around the ecliptic. The year of the seasons is shorter than the sidereal year by a little more than 20 minutes.

Tropic of Cancer.   The northern parallel of declination approximately $23°27'$ from the celestial equator, reached by the Sun at its maximum northern declination; or the corresponding parallel on the Earth.

Tropic of Capricorn.   The southern parallel of declination approximately $23°27'$ from the celestial equator, reached by the Sun at its maximum southern declination; or the corresponding parallel on the Earth.

tropopause.   In a planetary atmosphere, the boundary between the troposphere and stratosphere usually characterized by an abrupt change of temperature gradient (temperature change per kilometer of altitude). Atmospheric stability is greater above the tropopause; below it, the atmosphere is turbulent. Its height varies from 10 to 20 km on

Earth. In polar regions in winter it is often difficult or impossible to determine just where the tropopause lies, since under some conditions there is no abrupt change in lapse rate at any height.

troposphere.   The lowest layer of a planetary atmosphere in which the temperature decreases steadily with increasing altitude (Greek: *trope*, "turning.") It extends from Earth's surface to a height of about 10 to 20 km, depending on the latitude and time of year. Turbulence is greatest in this region, and most of the visible phenomena associated with the weather occur here; in this region nearly all clouds are formed.

trough.   On a weather map, an elongated area of relatively low pressure.

true meridian.   A great circle through a planet's poles of rotation, distinguished from magnetic meridian, grid meridian, etc.

true position.   The position of a celestial body (or space vehicle) on the celestial sphere as computed directly from the elements of the orbit of the Earth and the body concerned without allowance for light-travel time.

true prime vertical.   The vertical circle through the true east and west point of the horizon, as distinguished from magnetic, compass, or grid prime vertical through the magnetic compass or grid east and west points, respectively.

true sun.   The actual Sun as it appears in the sky, as distinguished from the mean Sun (q.v.).

Tucana (The Toucan).   A faint southern constellation, interesting because of several notable objects, including most of the Small Cloud of Magellan and, for the magnificent metal-rich globular cluster, 47 Tucanae, and several double stars. Tucana was introduced by Bayer in 1603. The total area is 295 sq degrees.

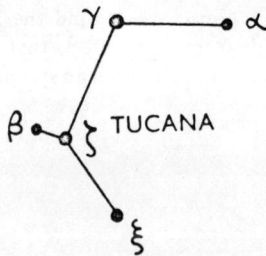

Tunguska event.   An immense atmospheric explosion over Siberia on June 30, 1908. Its energy was equivalent to a nuclear bomb blast of about 12 megatons, and it flattened trees 30 km away. The explosion is believed to be the result of atmospheric impact of a small comet nucleus or weakly consolidated meteorite that blew up before hitting the ground. Details of the observed trajectory and recent reports of microscopic carbonaceous chondrite meteorites in the soil support this view. Sensationalistic accounts have tried to assert that the explosion was the crash of a nuclear-powered spaceship.

turned edge.   The flattening of the edge of a telescope mirror from the paraboloidal curve.

21-cm radiation.   Emission from neutral atomic hydrogen atoms produced when the electron "flips," or reverses its spin from a direction parallel to the nucleus's spin to the opposite direction, which has lower energy. Since hydrogen is so abundant in space, the 21-cm radio radiation is very strong from massive interstellar gas clouds. It was first detected in 1951 and has been used to map interstellar gas and galactic structure.

twilight.   Sunlight diffused by the air onto a region of the Earth's surface where the Sun has already set or has not yet risen. Astronomical twilight ends in the evening and begins in the morning when the Sun's center is 18° below the horizon; the fainter stars are then visible overhead in a clear sky. The time of sunset and sunrise and the end and beginning of astronomical twilight can be found in some almanacs for any date and latitude.

Tycho's Nova (Cassiopeia).   In the year 1572 a new star appeared in the constellation of Cassiopeia, associated with Tycho Brahe the celebrated Danish astronomer. Tycho first saw it in November, and in his account said that it brightened rapidly from night to night until it became brighter than Sirius and Jupiter. The star then began to diminish in brightness until in March 1574 when it disappeared without a trace after having shown for a period of 17 months. An interesting thing about the star was its change in color. At its brightest it was dazzling white. Then as its brilliancy diminished, it changed to yellow, yellowish red, red, and finally to ashy paleness, and then extinction. The supernova remnant has been located as an x-ray source about 10,000 to 16,000 light-years away.

Tychonic system.   An obsolete theory of the solar system proposed by Tycho Brahe. In this system the Sun and Moon circle around the Earth, but the other planets revolved around the Sun. Aside from slight effects that could not have been detected without a telescope, the Tychonic and Copernican systems gave similar predictions of the positions of the planets.

U

U,B,V system. See photometry, three color.

U Geminorum stars. A common type of dwarf nova in which the light curves show no evidence for any period of constant intermediate brightness, thereby distinguishing them from the Z Camelopardalis variables. U Geminorum stars remain at minimum brightness for perhaps months on end, suddenly surging up for a brief maximum or rise, lasting for a few days at more than general brightness. The longer the period (which cannot be predicted), the greater the maximum. The range of these stars can be five magnitudes, or even more. None is ever visible with the naked eye.

ultraviolet light. Light of wavelength less than about 4000 Å, to which the eye is not sensitive. In 1802 the English physicist William H. Wollaston proved that there were radiations outside the violet end of the visible spectrum which, like visible light, are capable of darkening silver chloride. Although most of these radiations from celestial bodies are blocked by Earth's atmosphere, they have been studied by telescopes in satellites, rockets, and balloons.

ultraviolet stars. Very hot stars that radiate strongly in the ultraviolet. Many are small, dense stars, often in planetary nebulae (supernova remnants), apparently evolving by contraction into white dwarfs.

umbra. (1) The darkest central part of a shadow. When a source of light, not a point source, casts a shadow of an object the shadow consists of two parts: the umbra and the penumbra. The umbra or total shadow is the region completely cut off from the light while the penumbra is partly illuminated by some part of the light. (2) The darkest part of a sunspot.

Umbriel. A satellite of Uranus, believed to be roughly 1300 km in diameter.

uncertainty principle. The physical principle that a system (especially of subatomic particles) cannot be observed in great detail without disturbing the system (i.e., by shining light on it). The better the position of a particle is known, the more uncertain is its velocity, and vice versa. In particular the product of the minimum uncertainty in position times minimum uncertainty in momentum is of the order of Planck's constant, h. Sometimes called the Heisenberg uncertainty principle after the physicist who propounded it. Also called the indeterminacy principle.

Universal Time. In order to avoid mixups in the timing of a particular event, astronomers in the timing of a particular event record any instant of time as it would appear on a clock in Greenwich. Here the hours of the day are numbered from 0 to 24, using a four-digit number. Thus, 1315 U.T. is the same as 1:15 P.M. in Greenwich. Any local time may be changed to the time at Greenwich by adding a time correction. Thus, $U.T. = E.S.T. + 5h = P.S.T. + 8h.$

universe. Everthing that exists.

universe, expanding. See expanding universe.

universe, hyperbolic. A hypothetical model of the universe in which the galaxies are receding too fast to ever recombine. This requires no more than a limiting amount of absorbed intergalactic material since a greater amount would produce a gravitational attraction that could pull galaxies back. Existence of more than this amount of material is controversial, but now appears probable.

universe, steady state. See steady state theory.

upper air. In synoptic meteorology and in weather observing, that portion of the atmosphere which is above the lower troposphere.

upper atmosphere. The atmosphere above the stratosphere; beyond an altitude of about 50 km.

uranographer. An old term for one who specializes in making maps or pictures of the heavens. "Urano" comes from a Greek word meaning sky; "geo" comes from the name of the goddess Ge, goddess of the earth. "Grapher" comes from the word meaning drawing or writing.

Uranometria. A book published in 1603 in which Bayer introduced one of our present means of designating stars: the name of the constellation prefixed by a Greek letter. In general the stars are lettered in about their order of magnitude, α being the brightest, β the next brightest, and so on; but there are notable exceptions (i.e., the stars of the Dipper are lettered in order in which they occur). When Greek letters are exhausted, Roman letters are employed.

α	Alpha	ι	Iota	ρ	Rho
β	Beta	κ	Kappa	σ	Sigma
γ	Gamma	λ	Lambda	τ	Tau
δ	Delta	μ	Mu	υ	Upsilon
ϵ	Epsilon	ν	Nu	ϕ	Phi
ζ	Zeta	ξ	Xi	χ	Chi
η	Eta	o	Omicron	ψ	Psi
θ	Theta	π	Pi	ω	Omega

Uranus. Called Uranus after the god of the skies, which explains the significance of its sign ♅, an arrow pointing upward to the heavens. To the naked eye Uranus is barely visible in the clear moonless night, appearing as a star of sixth magnitude. It moves very slowly among the zodiacal stars, completing its circuit about the Sun in 84 years. Through a telescope the planet presents a tiny disk decidedly green in color. Vague belts have been suspected; and therefore it is believed that this planet resembles Jupiter in physical structure (though it is much smaller in size, with a diameter of about 32,000 miles). In March, 1977 as Uranus passed in front of a star, astronomers discovered a series of thin rings around Uranus, manifested as they occulted the star. There are five moons: Miranda, Ariel, Umbriel, Titania, Oberon. See also data in Appendix.

Uranus Rings.   At least nine narrow rings have been found since their first discovery in 1977. They lie 40,000 to 49,000 km from the center of Uranus ($\sim$1.6 to 2 planet radii), and are narrower and darker than Saturn's rings. They are probably confined in their orbits by unseen small satellites. They are composed of small particles and have estimated mass around 10^{19} g.

Ursa Major (The Great Bear).   The third in size of the constellations with an area of 1,280 sq degrees. It is an ancient and well-known star group to which Ptolemy assigned twenty-seven stars, and the figure made by the 10 brightest stars is familiar to every northern astronomer. The constellation contains many attractive double stars. There are no open clusters, but most of the bright stars are in a loose open cluster which is so close (68 light-years) as to cover much of our sky and be unrecognizable as a cluster. Although only one planetary nebula is known, extra galactic nebulae exist in profusion, many of them in groups.

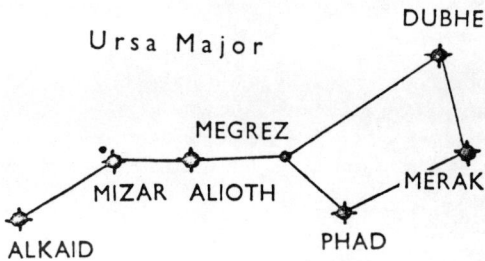

Ursa Major cluster.   See Ursa Major.

Ursa Minor (The Little Bear).   This small constellation (area 256 sq degrees) is an ancient star group to which Ptolemy assigned seven stars, but the origin of the name is uncertain. It is now distinguished by the proximity of the north pole of its leader Polaris, which makes this the most useful navigational star in the sky. Both this use and the name are comparatively modern, for in the time of Hipparchus (about 150 B.C.) the star was more than 12° distant from the pole. Closest approach will be 26'30" in A.D. 2095. Polaris itself is a wide unequal pair with rapidly changing R.A., as well as being a Cepheid variable of period 3.97 days and a spectroscopic binary of period 29.6 years. (See Fig., p. 240.)

U.V.   Common abbreviation for ultraviolet.

UVby colors.   See photometry, four-color.

V

Valles Marineris. The immense system of canyons on the equatorial region of Mars, discovered by Mariner 9 in 1971 and named for that spacecraft. Especially, the long linear canyon coincident with the stubby "canal," Coprates, seen in Earth-based telescopes.

Van Allen radiation belt. (1) J. A. Van Allen reported on May 1, 1958 (with data from the first American satellite, Explorer I) a major phenomenon in geophysics, the existence of a permanent toroidal (i.e., doughnut-shaped) belt (or belts) of high-energy radiation, probably consisting of protons and electrons surrounding Earth at an altitude above 650 to 800 km. These high energy particles are found at low and moderate (but not at high) latitudes.Several belts exist, consisting of ions temporarily trapped from the solar wind. Many eventually are "dumped" near the Earth's magnetic poles, creating aurorae as they strike the atmosphere. (2) By extension, similar belts around Jupiter or other planets with substantial magnetic fields.

van der Waals forces. Weak forces of attraction between adjacent neutral molecules or atoms, due to fluctuating asymmetries in their electron clouds.

vapor pressure. The pressure exerted by the vapor of a liquid (i.e., the atoms or molecules that escape from the liquid surface and create a gas above it). In meteorology, usually used to designate the vapor pressure of water.

vaporization. Conversion from solid or liquid into gas.

variable stars. Stars which are unstable, and fluctuate in brightness over a limited range. Some are regular (periodic) in behavior, others are erratic and unpredictable. The most important variables are probably the Cepheids, which are perfectly regular with periods of the order of a few days or weeks. Since their periods are directly related to their luminosities, they are very useful in allowing astronomers to measure luminosities, hence distances, thus estimating distances of clusters or galaxies in which they live. Cepheids are post-giant stars. Polaris is a Cepheid variable with the extremely small range of about 0.1 magnitude in blue light. The red giant long-period stars are also notable; an example is Mira in Cetus, which sometimes becomes as bright as Polaris and has a period of 331 days. There are many irregular variables such as the U Geminorum or SS Cygni stars which undergo periodic outbursts—intervals of a few weeks when they brighten by several magnitudes. Variability is associated with the post-giant star of the late stellar evolution. Also important are young, pre-main-sequence variable stars of which T Tauri variables (q.v.) are the most important. In the middle of the nineteenth century when the systematic study of these stars was begun, it was universally agreed that the capital Roman letters should be reserved for these stars, followed by the name of the constellation. Unfortunately the letters from A to Q had already been used for certain nonvariable stars in the newly charted southern constellations. For this reason the first variable to be discovered in any constellation is given the letter R followed by S, T, and so on to Z. When further letters are needed, the series recommences with RR, RS, RT, to RZ; SS, ST to SZ; TT, TU to TZ and so on, until ZZ. The nomenclature then continues with AA, AB, to AZ followed by BB, BC to BZ, finally ending with QZ.

variable stars, extrinsic.   Not true variable stars, as their changes of brightness are caused not by their own behavior but by the intervention of some external action such as variations in circumstellar dust, or by the changes of aspect, as when an ellipsoidal star revolves or rotates. The eclipsing variables, for example, change in brightness because two stars periodically eclipse one another, either totally or partially.

variable stars, secular.   A small number of the brighter stars are known which appear to have undergone changes in brightness over a period of many hundreds of years. The records upon which evidence of their variability is based are, for the most part, extremely scanty and often unreliable. Nevertheless a few of them have been found to exhibit small variations in brightness. The star Pleione in the Pleiades was estimated by Ptolemy to be comparable in brightness to its close neighbor Atlas, whereas now it is about two magnitudes fainter.

variation of latitude.   A small periodic change in the astronomical latitude of points on the Earth, due to wandering of the poles.

vector.   (1) A quantity which has both magnitude and direction. Force, acceleration, velocity, are vector quantities. (2) The arrow often used on diagrams to represent a vector. (3) As a verb, to direct an aircraft or spacecraft in a specified direction.

Vega.   See Alpha Lyrae.

Vela (The Sails).   One of the three large groups into which Gould in 1877 divided the ancient constellation Argo Navis. It lies north of Carina between Puppis and Centaures and is almost completely immersed in the Milky Way which in this region is cut by an irregular lobed band of dark absorbing material. The area of the constellation is 500 sq degrees and the center culminates at midnight about February 11th. There are many attractive double stars and 22 open clusters are known.

Vela pulsar.   Pulsar PSR 0833-4S, remnant of a supernova once visible (perhaps around 7000 B.C.) for a few thousand years. It lies at the center of the Gum nebula, a large cloud of expanding gas, and may be about 1300 to 1600 light-years away.

velocity.   A vector quantity measuring both speed and direction of a body, relative to some defined or assumed coordinate system.

velocity curve.   A plot of the variations in velocity as revealed by spectrum-line Doppler shifts as a function of time.

velocity of escape.   See escape velocity.

velocity of light.   186,000 miles per sec, or 3.0×10^{10} cm/sec, a constant for light as measured by all observers. Abbreviated c.

Venera.   Name of first series of space vehicles parachuted onto Venus by the Soviet Union. Venera 7 made the first successful spacecraft landing on any other planet in 1970. In 1975, Venera 9 made the first photos of the surface of another planet.

Venus.   The second planet from the Sun, at a mean distance of 108 million km. Because its orbit lies between Earth and the Sun, Venus telescopically goes through phases ranging from a narrow crescent (when between Earth and Sun) to a full disk (when on the far side of the Sun). Though its period of revolution around the Sun is 225 days, it repeats its cycle of phases in a synodic period of 584 days. In the telescope it appears as a featureless or nearly featureless disk covered with clouds. When seen as a narrow crescent, the horns of the crescent may be extended beyond a half circle because of light scattered in its atmosphere. Spectral measures show that the atmosphere is largely CO_2, a finding confirmed by several Russian and American Venus probes parachuted to the surface. After decades of speculation that Venus might have a swampy, watery surface, a dusty surface, or even an oily surface, these probes and Earth-based measurements established astonishing surface conditions: a temperature of 750 K (891°F), a surface atmospheric pressure about 90 times greater than Earth's, and virtually no water or water vapor. The atmosphere is about 97% CO_2 (by volume) with traces of N_2, CO, HCl, HF, and possibly H_2O, with clouds composed of tiny droplets of sulfuric acid. Photographs made at two landing sites on the surface showed a relatively clear scene (horizon visible) with scattered angular rocks at one site and eroded-looking, soil-sprinkled rocks at the other. Radar measures from Earth show large craters and canyon-like structures resembling rift or faulted valleys on Earth and Mars. Russian and American spacecraft in 1978 detected massive lightning storms near the surface, and measured hazy cloud tops 63–67 km above the surface, with the densest haze at 49–52 km, with the atmosphere below 33–49 km being clear of haze and dust.

Venus transits.   Transits of Venus are possible only within about a few days before or after June 7 and December 9, the dates when the Sun passes the nodes of the planet's path. They are less frequent because the limits are narrower and because conjunctions come less often. Transits of Venus now come in pairs having a separation of 8 years. The latest pair of transits occurred in 1874 and 1882, the next pair is due on June 8, 2004, and June 6, 2012. After a while there will be a long period when they occur singly. See transit.

vernal equinox.   The point where the Sun appears to cross the celestial equator from south to north; also called the First Point of Aries. Used as the origin of the coordinates right ascension and declination.

vernier.   A short scale arranged to slide along the graduations of the main indicating scale of an instrument. The vernier is graduated so that a certain number of its divisions are just equal to one more or less than the same number of divisions in the main scale. Fractions of divisions on the main scale are determined by observing which line on the vernier coincides with a line on the main scale, allowing greater accuracy than possible without the vernier.

vertex.   (1) The point on the limb of the Sun, Moon or a planet which is highest above the horizon. (2) (Rare) The zenith.

vertical circle.   An arc of a great circle drawn from the zenith through a star and perpendicular to the horizon.

very large array (VLA) telescope.   A radio telescope to be built in New Mexico, consisting of an array of 82-foot radio dish antennas, distributed at 27 points along three 13-mile lines. It is designed to give radio resolution comparable to the largest optical telescopes.

Vesta.   The brightest asteroid and fourth to be discovered (in 1807). It is not the biggest, being about 540 km across, half the size of Ceres; but it has an unusually high albedo, reflecting about 24% of the light striking it. Vesta's surface appears to resemble basaltic achondrite meteorites, based on spectroscopic meteorites.

vignetting.   Darkening of the corners of an image due to reduced light reaching points far off the optical axis.

virga.   Streamers of precipitation falling from a cloud which do not reach the ground.

Virgo (The Virgin).   The star group known as Virgo, the Virgin lies east of Leo, along the ecliptic. An imaginary line following the curve of the Big Dipper's handle helps locate Virgo by pointing first to the bright star, Arcturus, and then to Virgo's brilliant star, Spica. Virgo is a long constellation occupying a space of about 45 degrees along the Ecliptic; its stars roughly outline a human figure. On the ancient star maps Virgo is generally represented as a beautiful, stately woman with folded wings and in a walking attitude. In her left hand she holds a head of wheat or sheaf of grain, which is marked by Spica. The head of the figure is toward Regulus, marked by a compact group of faint stars, and the feet are represented by two stars south of Arcturus. Three stars are in her shoulder, one is in her extended right hand, and the others form her drapery. The constellation of Virgo is in a region of hundreds of distant galaxies, all beyond the reach of the naked eye. Virgo is the sixth of the zodiacal constellations with an area of 1,294 sq degrees.

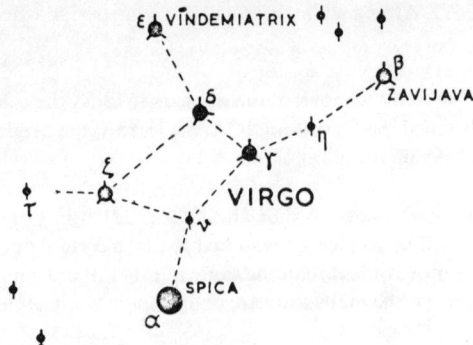

Virgo cluster.   A relatively nearby and prominent cluster of galaxies, of which some 2500 have been counted. It is located about 19 million parsecs away from the Milky Way in the constellation Virgo.

virtual temperature.   The temperature of damp air at which dry air of the same pressure would have the same density as the damp air. It is always higher than the actual temperature.

visual binary.   A physical double star in which both components can be resolved by the eye at the telescope.

visual magnitude.   The brightness of a star (expressed in the magnitude system) as estimated by the eye or as measured in wavelengths to which the eye is sensitive.

visual photometer.   An instrument attached to the telescope for measuring brightness visually.

VLA.   See very large array telescope.

VLB.   Very long baseline radio interferometry.

Volans (The Flying Fish).   One of Bayer's southern additions in 1603. It follows the large Magellanic Cloud between it and Carina. The area is 141 sq degrees. Apart from a few double stars, the only objects of telescopic interest for small apertures are extragalactic objects.

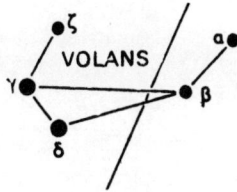

volatiles.   Elements or chemical compounds easily driven off by heating of planetary or meteoritic matter, such as hydrogen, water, mercury, etc. They tend to have low melting or boiling points. See also refractories.

volvelle.   A medieval instrument for illustrating the phases of the Moon and its positions in the sky relative to the Sun and the Earth.

Vulcan.   A sizable hypothetical planet once suspected to exist between the Sun and Mercury. Although reported by a few observers in the 1800's, its existence is now discounted.

Vulcanoids.   Small asteroids hypothesized by some dynamicists possibly to lie between the Sun and Mercury. None had been discovered through 1978 in spite of searches.

Vulpecula (The Fox).   A small obscure constellation which is not ancient. Vulpecula, formerly Vulpecula et Anser, the Fox and Goose, was created by Hevelius. It lies between Sagitta and Delphinus to the one side and Cygnus to the other, but will not easily be identified as it has no stars as bright as the fourth magnitude. There is one remarkable object, the so-called Dumb bell Nebula, well below naked-eye visibility. The area is 268 sq degrees. There are 9 substantial open star clusters, and 3 planetary nebulae. The screening effect of the Milky Way practically excludes distant galaxies from this constellation.

VV Cephei.   A variable star with the exceptionally long period of 7,430 days. It is one of a small group of Algol variables having these remarkably long periods. VV Cephei consists of two components which differ in size and spectral type. The masses have been estimated as 30 and 50 times that of the Sun with spectral classes B_9 and M_3, respectively. The apparent visual magnitude of the system at maximum is 6.5 and the change in brightness, measured photoelectrically is $0^m.8$.

W

W cloud. A cloudy marking shaped like the letter W that occasionally appears on blue photographs of Mars when the image is viewed with the south pole uppermost. Mariner 9 in 1971 to 1972 revealed that the W is caused by clouds forming over a group of high Martian volcanoes.

W Ursae Majoris stars. A group of short-period eclipsing stars consisting of pairs which are small and close together or in contact. Named after W Ursae Majoris, they are the first of this type to be discovered. The period of W UMa is only 8 hours and the surfaces of the stars are nearly touching. They are thought to be embedded in mutual nebulosity. The two stars are nearly equal in size and brightness, but not in mass, and there is an exchange of material between the two.

W Virginis star. A Cepheid of Population II.

walled plain. Obsolete term for large lunar craters.

wandering of the poles. The shifting rotation axis of the Earth with respect to the physical globe caused in part by mass redistribution in the interior by magma flow, convection, earthquakes, etc.

wane. To grow smaller or phase, as phase angle increases; opposite of wax.

warm front. The advancing edge of a warm mass of air.

waterspout. A whirlwind formed on hot days over water, in which water droplets and cloud particles are carried some distance (from a few feet to several hundred feet) into the air. In rare cases a tornado may appear over the water and give rise to an exceptionally severe form of waterspout.

watt. A standard unit of power equaling 10^7 ergs/sec and often used in measuring radio signal strength.

wave front. A surface defined by an advancing wave. Every point on the front is in the same phase or state of vibration.

wavelength. The distance that light (or any other signal propagated by waves) travels in one complete vibration of the source. The distance between the crests (or troughs) of regular waves. Visible light of various colors has wavelengths ranging from about 1/10,000 inch (4000 A) to about twice that length. See also spectrum.

wave-theory of light. A theory of light clearly enunciated by Thomas Young and Augustin Fresnel early in the 19th century. According to it, light is a transverse wave motion in hypothetical medium called the "Luminiferous Ether." The ether does not exist, but light does have some properties of waves, as well as some properties of particles.

wax.   To increase in illuminated portion, as the phase angle decreases from 180° to 0°.

weather.   The state of the atmosphere at a given moment with respect to temperature, moisture, cloudiness, precipitation, or other meteorological phenomena.

wedge.   (1) On an isobaric weather map, an elongated area of relatively high pressure. (2) A glass prism.

weight.   A measure of the gravitational force of attraction on a body usually at the surface of a planet which tells us how heavy it is. Weight should not be confused with mass. The mass of a body is constant. The weight, however, decreases as the body moves away from the planet, as weight is the measure of the gravitational attraction of any one mass for another.

wet bulb.   The bulb of a thermometer which is kept moistened when taking humidity measurements.

Whirlpool Galaxy (Bootes).   A great spiral galaxy, M51 (NGC 5194), more condensed than the nebula in Andromeda but of the same general character. It is located at a distance of about 12 million light-years or more.

whistlers.   Radio waves generated by lightening and traveling large distance along the Earth's magnetic field.

whistling meteor.   Name applied to a radio meteor when a detection system is used in which the presence of the meteor is indicated by a rapidly changing audio frequency radio signal.

white dwarfs.   Earth-sized hot dense stars, usually faint white to blue white, believed to constitute several percent of the total stars in the Galaxy. They are about 10 magnitudes fainter than stars of the same color (or spectral class) in the main sequence; the radii are smaller by a factor of approximately a hundred. A white dwarf in the same spectral class as the Sun would have a radius similar to Earth. The masses of the white dwarfs are of the same order as the Sun, therefore they are extremely dense bodies. Because of their large densities and surface gravitational forces, the outer layers (or atmospheres) are also very dense and have high gas pressures, manifested by the marked pressure broadening of the spectral lines of the white dwarfs. They are stars in a collapsed state, created when the interior energy sources needed to keep the gas expanded are exhausted and gravity collapses the star to densities controlled by the pressure exerted by degenerate electrons.

Widmanstatten figure.   The pattern of intersecting mineral bands which appears when iron meteorites are cut, polished, and etched with dilute acid, caused by interlocking growth of iron crystals. Terrestrial iron lacks this feature, which requires long cooling time deep in the interior of a parent planetary body or asteroid, from which meteorites are derived.

width of a spectral line.   Wavelength interval covered by a spectral line. Neither absorption lines nor emission lines are at exactly one wavelength; many lines in optical spectra have a spread of several angstroms.

Wien's Law.   A law giving the wavelength (color) at which most light is emitted for any radiating body of specified temperature. In general, the hotter a body, the shorter the wavelength (bluer the color) of its emitted light.

Wilson effect.   First noticed in 1769 when Alexander Wilson, Professor of astronomy at Glasgow, followed a large sunspot as it approached the western limb. He noticed that as it neared the limb, the umbra steadily became more and more displaced toward the center of the Sun. Wilson realized the significance of this. The spot must have been shallow with the umbra lying below the level of the photosphere, rather like a saucer that is viewed from a more and more sharply inclined angle. His supposition was confirmed when the spot, which proved to be a long-lined one, reappeared at the eastern limb with its umbra once more displaced, becoming central as it moved toward the center of the disk.

window.   A wavelength interval in which an atmosphere (usually Earth's) is nearly transparent to light.

Wind symbols.   Wind symbols with Beaufort scale and description—

BEAU-FORT NUMBER	MAP SYMBOL	DESCRIPTIVE WORD	VELOCITY (MILES PER HOUR)
0	⊙	CALM	LESS THAN 1
1			1 TO 3
2		LIGHT	4 TO 7
3		GENTLE	8 TO 12
4		MODERATE	13 TO 18
5		FRESH	19 TO 24
6			25 TO 31
7		STRONG	32 TO 38
8			39 TO 46
9		GALE	47 TO 54
10			55 TO 63
11		WHOLE GALE	64 TO 75
12		HURRICANE	ABOVE 75

winter solstice.   The date (about December 22) when the Sun reaches its greatest distance south of the equator. It is the first day of winter.

Wolf-Rayet stars.   Hot stars losing mass in the form of expanding gas envelopes. These stars are named after two astronomers at the Paris Observatory who, in 1867, discovered the first known stars of this remarkable type. About 200 are now recognized; the brightest is the second magnitude northern star Gamma Velorum. They are sometimes grouped in the spectral classification sequence as type W. Of average absolute magnitude -5, they are among the hottest stars, having surface temperatures as high as 50,000° K. Averaging twice the Sun's diameter, they are surrounded by much larger atmospheres.

world line.   An imaginary line through space and time recording the positions of a particle at all moments during its history.

X

x-rays. Radiations of wavelengths shorter than those in the ultraviolet, ranging in wavelength from about 0.1 to 100 Å. They can be readily produced in the laboratory by bombarding a metal in vacuum with a stream of high velocity electrons. X-rays from the Sun are entirely absorbed in the Earth's atmosphere.

x-ray sources. Starlike objects strongly emitting x-rays. They are all members of binary star systems and fluctuate in brightness on timescales ranging from fractions of a second to several weeks. They are believed to involve streaming of gas off one member of the binary pair, after which the gas crashes onto the other star's surface causing high temperatures and x-ray emission.

x-ray telescope. A set of x-ray detectors stacked in such a way as to count x-rays only from a single direction.

Y

year. An interval of time based upon the revolution of the Earth in its orbit around the Sun. See tropical year and sidereal year.

yellow clouds. Martian clouds caused by blowing dust.

yoke mounting. A form of equatorial telescope mounting in which the polar axis consists of a yoke in which the telescope tube is mounted.

Z

z.   Abbreviation for red shift, defined as (shift in wavelength)/(original wavelength).

Z Camelopardalis stars.   A small subgroup of variable stars, similar to prototype Z Camelpardalis, which varies between 10.2 and 13.4 magnitudes in a mean period of 20 days. In their general behavior they are very much like the U Geminorum stars, subject to similar nova-like outbursts but having shorter mean periods, spending less time at minimum and having smaller amplitudes. The major difference, which justifies their inclusion in a separate subgroup, is the periods of "standstill": following certain maxima, the brightness does not fade to a normal minimum but remains virtually constant at some intermediate magnitude.

ZAMS.   Abbreviation for zero age main sequence, the main sequence as defined by stars of zero age (i.e., prior to substantial evolution).

Zeeman effect.   The splitting of spectral lines into two or three polarized components when the atoms causing the lines are in a magnetic field. The separation of the components depends upon the type of atom and the magnetic field strength. The polarization depends on the viewer's direction relative to the magnetic field. It can be used to measure the magnetic strength in the gas in which the line arises.

zenith.   The point vertically overhead on the celestial sphere. The zenith is in the opposite direction from the nadir, the point directly underfoot.

zenith distance.   Angular distance from the zenith.

zenographic.   Referring to positions on Jupiter measured in latitude from Jupiter's equator and in longitude from a reference meridian.

zero age main sequence.   See ZAMS.

zodiac.   The zodiac is an imaginary band in the sky roughly 11° wide in the center of which lies the ecliptic. It is within this band that the Sun, Moon, and planets travel. The ancients defined 12 constellations along this band, and referred to them as the Signs of the Zodiac. The signs bear the same names as the constellations. They are: Aries, Taurus, Gemini, Cancer, Leo, Virgo, Libra, Scorpius, Sagittarius, Capricornus, Aquarius and Pices. The signs were defined in sequence starting from the vernal equinox. Because of the precession of the equinoxes, the vernal equinox has shifted the sign positions with respect to the stars, away from the constellations after which they were named. Since the signs are counted from the vernal equinox, the signs and constellations of the zodiac of the same names no longer have the same positions. Thus, the signs (used by astrologers in casting horoscopes) do not correspond to the real star patterns (See Fig., p. 254.)

ARIES The Ram	♈	TAURUS The Bull	♉	GEMINI The Twins	♊
CANCER The Crab	♋	LEO The Lion	♌	VIRGO The Virgin	♍
LIBRA The SCALES	♎	SCORPIUS The Scorpion	♏	SAGITTARIUS The Archer	♐
CAPRICORNUS The Goat	♑	AQUARIUS The Water-Bearer	♒	PISCES The Fishes	♓

Signs of the Zodiac

zodiacal constellations. The twelve constellations that lie along the zodiac, containing the apparent path of the Sun, Moon, and planets.

zodiacal light. A name given to an elongated triangular patch of light located near the ecliptic, on the western horizon after sunset, and on the eastern horizon before dawn. It is sunlight reflected from small meteoritic particles or dust in the same plane as the planetary orbits.

zone (telescope). A doughnut-shaped area of the mirror, measured from the center.

zone of avoidance. An irregular band along the Milky Way in which no distant galaxies can be seen due to obscuration by the interstellar gas and dust, lying in the plane of our galaxy.

zone time. Standard time used in a time zone or band of longitude roughly 15 degrees wide.

Appendix

Table 1. Metric and Astronomical Measurements

THE METRIC SYSTEM

General System of Multiples

Multiple	Prefix	Symbol
10^{12}	tera	T
10^{9}	giga	G
10^{6}	mega	M
10^{3}	kilo	k
10^{2}	hecto	h
10	deka	da
10^{-1}	deci	d
10^{-2}	centi	c
10^{-3}	milli	m
10^{-6}	micro	μ
10^{-9}	nano	n
10^{-12}	pico	p
10^{-15}	femto	f
10^{-18}	atto	a

LENGTH

Metric to English

1 Ångstrom = 1Å = 10^{-8} cm (centimeter).
1 micron = 10^{-3} mm (millimeter) = 10^{-6} m (meter).
1 cm = 0.3937 in. (inch).
1 m = 3.281 ft. (foot).
1 km (kilometer) = 0.62140 mi. (mile).
1 light-year = 9.461×10^{12} km = 5.88×10^{12} mi.

English to Metric

1 in. = 2.540 cm.
1 ft. = 30.48 cm.
1 yd. (yard) = 0.9144 m.
1 mi. = 1.609 km.

VOLUME

Metric to English

1 cm^3 = 0.0610 cu. in.
1 liter = 0.2643 U.S. gal. (gallon).
1 liter = 0.220 Imperial gal.

English to Metric

1 cu. in. = 16.387 cm^3.
1 U.S. gal. = 3.785 liters.
1 Imperial gal. = 4.546 liters.

WEIGHT

Metric to English (Avoirdupois)

1 g (gram) = 0.03527 oz. (ounce).
1 kg = 2.205 lb. (pound).
1 metric ton = 1,000 kg = 0.9842 long tons = 1.1032 short tons.

English (Avoirdupois) to Metric

1 oz. = 28.35 g.
1 lb. = 16 oz. = 0.4536 kg.
1 long ton = 1.016 metric tons.

TEMPERATURE

Metric to English: Degrees Fahrenheit = °F = (°C $\times$ 9/5) + 32.
English to Metric: Degrees Celsius = °C = (°F − 32) $\times$ 5/9.
Celsius to Kelvin (Celsius Absolute): K = °C + 273.2.
[Absolute zero, the temperature at which all molecular thermal motion ceases, is −273.18°C = −459.72°F = 0 K.]

ASTRONOMICAL MEASUREMENTS

1 astronomical unit	= 149.5 million km = approx. 93 million miles	
1 light-year	= 9.46×10^{12} km	
	= approx. 6 million million miles	
1 parsec	= 3.08×10^{13} km	
	= approx. 20 million million miles	

255

Table 2. Terrestrial Data

Polar diameter	12,714 km (7900 miles)
Equatorial diameter	12,756 km (7926 miles)
Mean diameter	12,742 km (7918 miles)
Oblateness (fractional degree of flattening)	$1/298 = 0.003353$
1° of latitude at poles	111.7 km (69.4 miles)
1° of latitude at equator	110.6 km (68.7 miles)
Land area	149,000,000 sq. km (29% of total)
Water area	361,000,000 sq. km (71% of total)
Total area	510,000,000 sq. km
Volume	11×10^{11} cu. km.
Mass	6×10^{27} grams
Density	5.5 gm/cm^3 (water $= 1$)
Nautical mile (minute of arc along a great circle)	6,080 feet
	1.15 statute miles
	1.85 kilometers
Statute mile	5,280 feet
	0.87 nautical mile
	1.61 kilometers

Dip of horizon in minutes of arc equals approximately the square root of observer's height in feet above sea level. (Example: Dip = 6 minutes of arc for height of 36 feet)

Distance of sea horizon in statute miles equals approximately 1.3 times the square root of observer's height in feet above sea level. (Example: Distance = 130 miles for height of 10,000 feet)

Curvature of earth in feet equals about 2/3 of the square distance in miles. (Example: Curvature = 6 feet at a distance of 3 miles)

Atmospheric pressure at sea level equals 15 pounds per square inch, supporting a column of mercury 30 inches (76 centimeters) high.

Atmospheric pressure diminishes about one-half for each 3.5 miles (5.6 kilometers). (Examples: Pressure at 7 miles = 7 1/2 inches; pressure at 10.5 miles = 3 3/4 inches)

Greatest height of aurora = 680 miles (1,100 km)

Atmospheric refraction = 35 minutes of arc at horizon, 5 minutes of arc at altitude of 10°, 1 minute of arc at 45°, 0 at zenith.

Albedo (reflecting power) or earth	about 0.4, depends on cloud cover
Velocity of escape	7 mi./sec. (11km/sec)
Period of earth's rotation (sidereal)	23^{h}56^{m}4^s.09
Inclination of equator to orbit	23°27′
Orbital velocity	29.8 km per second
Eccentricity of orbit	0.01674
Precessional cycle	25,800 years

Equinoxes move westward at rate of 50″.3 per year

Table 3. Lunar Data

Distance of Moon from Earth	
Greatest	406,697 km (252,710 miles)
Least	356,700 km (221,643 miles)
Mean	384,403 km (238,857 miles)
Equatorial horizontal parallax at mean distance: 57′03″	
Apparent angular diameter:	
Minimum	29′21″
Maximum	33′30″
Mean	31′05″
Eccentricity of orbit	1/18
Diameter of Moon	3,476 km (2,160 miles)
Volume	1/49 that of Earth
Mass	1/81 that of Earth
Mean density	3/5 that of Earth
	(3.34 g/cm^3)
Surface gravity	1/6 that of Earth
Velocity of escape	1 1/2 miles per second
Approximate temperature of soil	
At noon	100°C 212°F
At midnight	−150°C −238°F
Revolution and Rotation	
Synodic month (from one new moon to the next)	29^d 12^h 44^m 2.8
Sidereal month (true period of revolution around Earth)	27^d 7^h 43^m 11.5
Period of axial rotation	27^d 7^h 43^m 11.5
Inclination of orbit to ecliptic	5°08′
Period of revolution of nodes	18.6 years
Daily retardation in crossing meridian	
Minimum	38 minutes
Maximum	66 minutes
Average	50 1/2 minutes
Average velocity of Moon around Earth	
Linear	2,287 miles an hour
	or 3,350 feet a second
Angular	13°.2 a day or 33′ an hour
(Moon moves in one hour a distance about equal to its own diameter)	
Age of moon	about 4.5 billion years
	(= 4,500,000,000 years)
Age of major lava flows	about 3.5 billion years

Table 4. Major Objects in the Solar System
(Including sun, all planets and satellites known through 1978, and the five largest asteroids)

Object	Equatorial Diameter (km)	Rotation Period (days)	Distance from Primary (10^3 km unless marked)	Orbit Inclination[c] (degrees)	Orbit Eccentricity
Sun	1,392,000	25.4	0	—	—
Mercury	4,880	58.6	.387 AU	7°.0	.206
Venus	12,104	243 R[a]	.723 AU	3.4	.007
Earth	12,756	1.00	1.00 AU	0.0	.017
Moon	3,476	27.3	384	18–29	.055
Mars	6,787	1.02	1.52 AU	1.8	.093
Phobos	22	0.32	9	1.1	.021
Deimos	12	1.26	23	1.6	.003
Asteroids					
1[b] Ceres	1,003	0.38	2.77 AU	10.6	.08
2 Pallas	608	0.4?	2.77 AU	34.8	.24
4 Vesta	538	0.22	2.36 AU	7.1	.09
10 Hygiea	450	?	3.15 AU	3.8	.10
31 Euphrosyne	370	?	3.16 AU	26.3	.22
Jupiter	142,800	0.41	5.20 AU	1.3	.048
5 Amalthea	240	0.4?	181	0.4	.003
1 Io	3,640	1.77	422	0.0	.000
2 Europa	3,050	3.55	671	0.5	.000
3 Ganymede	5,270	7.16	1,070	0.2	.001
4 Callisto	5,000	16.69	1,880	0.2	.01
13 Leda	8?	?	11,110	26.7	.146
6 Himalia	170?	?	11,470	27.6	.158
10 Lysithea	19?	?	11,710	29.0	.130
7 Elara	60?	?	11,740	24.8	.207
12 Ananke	17?	?	20,700	147 R[a]	.17
11 Carme	24?	?	22,350	164 R[a]	.21
8 Pasiphae	27?	?	23,300	145 R[a]	.38
9 Sinope	21?	?	23,700	153 R[a]	.28
Saturn	120,000	0.43	9.54 AU	2.49	.056
11	300?	?	151?	0?	.0?
10 Janus	220?	?	160	0	.0
1 Mimas	400?	?	186	1.5	.02
2 Enceladus	500?	1.37	238	0.0	.00
3 Tethys	1,000?	?	295	1.1	.00
4 Dione	1,150	2.7	377	0.0	.00
5 Rhea	1,600	4.4	527	0.4	.00
6 Titan	5,000	15.95	1,222	0.3	.03
7 Hyperion	500?	?	1,481	0.4	.10
8 Iapetus	1,800	79.33	3,560	14.7	.03
9 Phoebe	200?	?	12,930	150 R[a]	.16

Table 4 (*Continued*)
(Including sun, all planets and satellites known through 1978, and the five largest asteroids)

Object	Equatorial Diameter (km)	Rotation Period (days)	Distance from Primary (10³ km unless marked)	Orbit Inclination[c] (degrees)	Orbit Eccentricity
Uranus	51,800	0.96? R[a]	19.18 AU	0.8	.05
5 Miranda	800?	?	130	3.4	.02
1 Ariel	2,000?	?	192	0	.00
2 Umbriel	1,300?	?	267	0	.00
3 Titania	2,400?	?	438	0	.00
4 Oberon	2,200?	?	586	0	.00
Neptune	49,000	0.92?	30.07 AU	1.8	.01
1 Triton	5,000?	5.9	354	160.0 R[a]	.00
2 Nereid	700	?	5,570	27.5	.76
Pluto	2500?	6.4	39.44 AU	17.2	.25
1 To be named	1000?	6.4	17	0	.0

[a]R in column 3 indicates retrograde rotation; in column 5, retrograde revolution.
[b]Numbers assigned to asteroids and outer planets' satellites indicate order of discovery, except for largest satellites, which are numbered arbitrarily.
[c]This column gives inclination to ecliptic for planets; to planets' equator for satellites.
Ref:
Hartmann, W. K. (1978) Astronomy: The Cosmic Journey (Belmont, CA: Wadsworth).
Morrison, D., and D. P. Cruikshank (1974), Space Science Review, *15*, 641.

Table 5. Asteroids and Related Interplanetary Bodies
(Giving properties of typical small bodies in the solar system)

Name	Semimajor Axis (Ave. distance from sun, in astronomical units)	Estimated Diameter (km)	Probable Composition (Usually based on spectroscopic data)
First 20 numbered asteroids in the main asteroid belt			
1 Ceres	2.77	1017	Carbonaceous chondrite
2 Pallas	2.77	585	Carbonaceous chondrite
3 Juno	2.67	247	Stony-iron
4 Vesta	2.36	531	Basaltic achondrite
5 Astraea	2.58	122	Stony-iron
6 Hebe	2.43	204	Stony-iron
7 Iris	2.39	205	Stony-iron
8 Flora	2.01	161	Stony-iron
9 Metis	2.39	160	Stony-iron
10 Hygiea	3.15	450	Carbonaceous chondrite

Table 5 (*Continued*)
(Giving properties of typical small bodies in the solar system)

Name	Semimajor Axis (Ave. distance from sun, in astronomical units)	Estimated Diameter (km)	Probable Composition (Usually based on spectroscopic data)
11 Parthenope	2.45	150	Stony-iron
12 Victoria	2.34	136	Stony-iron
13 Egeria	2.58	243	Carbonaceous chondrite
14 Irene	2.59	152	Stony-iron
15 Eunomia	2.64	260	Stony-iron
16 Psyche	2.93	250	Metallic or enstatite-rich
17 Thetis	2.47	98	Stony-iron
18 Melpomene	2.30	161	Stony-iron
19 Fortuna	2.44	221	Carbonaceous chondrite
20 Massalia	2.41	133	Stony-iron

Selected Mars-crossing Asteroids (*Often called Amor asteroids*)

Asteroids that cross Mars' orbit but don't get in as far as earth's orbit.

433 Eros	1.46	$7 \times 19 \times 30$	Stony-iron
887 Alinda	2.52	6	Stony-iron
1221 Amor	1.92	?	?
1580 Betulia	2.20	?	?
1915 Quetazlcoatl	2.52	?	?

Selected Earth-crossing Asteroids (*Often called Apollo asteroids*)

Asteroids that pass inside earth's orbit

1566 Icarus	1.08	1	Chondritic meteorite
1620 Geographos	1.24	3	Stony-iron
1685 Toro	1.37	3	Chondritic meteorite
1862 Apollo	1.47	?	?
(1976 AA)	0.98	1	Stony-iron

Selected Trojan Asteroids

Asteroids in Jupiter's orbit (P = preceding Jupiter by 60°; F = following Jupiter by 60°)

617 Patroclus (F)	5.21	147	Dark-colored rock
624 Hektor (P)	5.12	100×300	Dark-colored rock
1172 Aeneas (F)	5.17	130	Dark-colored rock
1173 Anchises (F)	5.17	92	Dark-colored rock

Remote asteroid

944 Hidalgo	5.82	20?	Rocky with some ice??

Table 5 (*Continued*)
(Giving properties of typical small bodies in the solar system)

Name	Semimajor Axis (Ave. distance from sun, in astronomical units)	Estimated Diameter (km)	Probable Composition (Usually based on spectroscopic data)
Remote asteroid-like object			
Chiron	13.70	100 to 650?	Rocky or icy??
Short-period comets			
Encke	2.2	2?	Icy
Halley	18	1 to 10?	Icy
Long-period comet			
Kohoutek	1000's	1 to 10?	Icy
Meteorites			
Pribram	2.46	fragment	Chondrite
Lost City	1.66	fragment	Chondrite
Meteors and fireballs			
1966 Perseid	40.2	fragment	dust particle?
1965 Leonid	52.6	fragment	dust particle?
1965 Draconid	3.5	fragment	dust particle?
Dec., 1966 Fireball	2.1	fragment	dust particle?
Jul., 1966 Fireball	32.3	fragment	dust particle?

Data taken from:

Morrison, D. (1977), Icarus, *31*, 185.

Chapman, C. R., J. Williams, W. K. Hartmann (1978) Annual Review of Astron. and Astrophys., *16*, 33.

Hartmann, W. K. (1978) Astronomy: The Cosmic Journey (Belmont, CA: Wadsworth).

Table 6. The Constellations

Name	Genitive ending (used to refer to stars, as in Alpha Andromedae)	Meaning	Abbre- viation
Andromeda	-dae	Chained maiden	And
Antlia	-liae	Air pump	Ant
Apus	-podis	Bird of paradise	Aps
Aquarius	-rii	Water bearer	Aqr
Aquila	-lae	Eagle	Aql
Ara	-rae	Altar	Ara
Aries	-ietis	Ram	Ari
Auriga	-gae	Charioteer	Aur
Boötes	-tis	Herdsman	Boo
Caelum	-aeli	Chisel	Cae
Camelopadus	-di	Giraffe	Cam
Cancer	-cri	Crab	Cnc
Canes Venatici	-num -corum	Hunting dogs	CVn
Canis Major	-is -ris	Great dog	CMa
Canis Minor	-is -ris	Small dog	CMi
Capricornus	-ni	Sea goat	Cap
Carina	-nae	Keel	Car
Cassiopeia	-peiae	Lady in chair	Cas
Centaurus	-ri	Centaur	Cen
Cepheus	-phei	King	Cep
Cetus	-ti	Whale	Cet
Chamaeleon	-ntis	Chamaeleon	Cha
Circinus	-ni	Compasses	Cir
Columba	-bae	Dove	Col
Coma Berenices	-mae -cis	Berenice's hair	Com
Corona Australis	-nae -lis	S crown	CrA
Corona Borealis	-nae -lis	N crown	CrB
Corvus	-vi	Crow	Crv
Crater	-eris	Cup	Crt
Crux	-ucis	S cross	Cru
Cygnus	-gni	Swan	Cyg
Delphinus	-ni	Dolphin	Del
Dorado	-dus	Dorado fish	Dor
Draco	-onis	Dragon	Dra
Equuleus	-lei	Small horse	Equ
Eridanus	-ni	River Eridanus	Eri
Fornax	-acis	Furnace	For
Gemini	-norum	Heavenly twins	Gem
Grus	-ruis	Crane	Gru
Hercules	-lis	Kneeling giant	Her
Horologium	-gii	Clock	Hor
Hydra	-drae	Water monster	Hya

Table 6 (*Continued*)

Name	Genitive ending (used to refer to stars, as in Alpha Andromedae)	Meaning	Abbre-viation
Hydrus	-dri	Sea-serpent	Hyi
Indus	-di	Indian	Ind
Lacerta	-tae	Lizard	Lac
Leo	-onis	Lion	Leo
Leo Minor	-onis -ris	Small Lion	LMi
Lepus	-poris	Hare	Lep
Libra	-rae	Scales	Lib
Lupus	-pi	Wolf	Lup
Lynx	-ncis	Lynx	Lyn
Lyra	-rae	Lyre	Lyr
Mensa	-sae	Table (mountain)	Men
Microscopium	-pii	Microscope	Mic
Monoceros	-rotis	Unicorn	Mon
Musca	-cae	Fly	Mus
Norma	-mae	Square (level)	Nor
Octans	-ntis	Octant	Oct
Ophiuchus	-chi	Serpent bearer	Oph
Orion	-nis	Hunter	Ori
Pavo	-vonis	Peacock	Pav
Pegasus	-si	Winged horse	Peg
Perseus	-sei	Champion	Per
Phoenix	-nisis	Phoenix	Phe
Pictor	-ris	Painter's easel	Pic
Pisces	-cium	Fishes	Psc
Piscis Austrinus	-is -ni	S fish	PsA
Puppis	-ppis	Poop (stern)	Pup
Pyxis	-xidis	Compass	Pyx
Reticulum	-li	Net	Ret
Sagitta	-tae	Arrow	Sge
Sagittarius	-rii	Archer	Sgr
Scorpius	-pii	Scorpion	Sco
Sculptor	-ris	Sculptor	Scl
Scutum	-ti	Shield	Sct
Serpens (Caput and Cauda)	-ntis	Serpent (Head and Tail)	Ser
Sextans	-ntis	Sextant	Sex
Taurus	-ri	Bull	Tau
Telescopium	-pii	Telescope	Tel
Triangulum	-li	Triangle	Tri
Triangulum Australe	-li -lis	S Triangle	TrA
Tucana	-nae	Toucan	Tuc
Ursa Major	-sae -ris	Great Bear	UMa
Ursa Minor	-sae -ris	Small Bear	UMi

Table 6 (*Continued*)

Name	Genitive ending (used to refer to stars, as in Alpha Andromedae)	Meaning	Abbre-viation
Vela	-lorum	Sails	Vel
Virgo	-ginis	Virgin	Vir
Volans	-ntis	Flying fish	Vol
Vulpecula	-lae	Small fox	Vul

Data adapted from C. W. Allen (1973) Astrophysical Quantities (London: Athlone Press).

Table 7. Notes on Selected Later Constellations added in Historic Times
(Often named after shipboard instruments of early explorers.)

1. Antlia—Originally Antlia pneumatica, the Air Pump
2. Caelum—The Chisel, originally Caelum sculptoris, the Sculptor's Chisel
3. Circinus—The Compasses, originally Circinus et norma, Compasses and Ruler
4. Columba—The Dove suggested by the Dutch geographer Pieter Plancius
5. Coma Berenices—The Hair of Berenice, suggested by the Greek astronomer Hipparchus, but not accepted by Ptolemy
6. Crux—The Southern Cross
7. Dorado—The Swordfish, or the Gold fish
8. Fornax—The Furnace, originally Fornax Chemica
9. Horologium—The Pendulum Clock
10. Lynx—The name was introduced by Hevelius of Danzig in the seventeenth century because the area which had to be filled in by this constellation is devoid of bright stars and therefore Hevelius said that a man needed lynx eyes to see anything at all.
11. Mensa—The Table; originally Mons Mensa, Table Mountain, meaning the mountain near Cape Town from which the French astronomer Nicolas Louis de Lacaille, who introduced this constellation made his observations of the southern sky.
12. Microscopium—The Microscope
13. Musca—The Fly; originally Musca australis, the southern fly.
14. Norma—The Ruler (in the sense of straight edge)
15. Octans—The Octant, originally Octans Hadleianus, Hadley's Octant.
16. Pictor—The Painter, originally Equuleus pictoris, the Painter's Easel.
17. Pyxis—The Mariner's Compass; originally Pysix nautica (The word pyxis, which literally means a small wooden box was used for the original mariner's compass because it was a wooden bowl, filled with water, in which a piece of reed supporting the magnetic needle, floated.)
18. Reticulum—The Net; originally Lacaille's Reticule romboide, the instrument he used for measuring angular distances between stars.
19. Scutum—The Shield, originally Scutum Sobiesii, named by Havelius in honor of the Plish king Jan Sobieski.
20. Sextans—The Sextant
21. Telescopium—The Telescope
22. Triangulum australis—The Southern Triangle
23. Tucana—The Toucan (as a typical bird of the southern hemisphere)
24. Volans—Originally Piscis volans, the Flying Fish.
25. Vulpecula—The Fox; the original name by Johannes Hevelius was Vulpecula cum ansere, the Little Fox with the Goose.

Table 8. The 24 Nearest Stars (out to 4 parsecs)
(In order of distance)

Name	Component (If multiple)	Apparent Visual Magnitude	Distance (parsecs)	Spectral Type	Semi-Major Axis in Multiple Systems (AU)
Sun	A	-27	0.0	G Main seq.	5.2
(Jupiter)	B	—		(Planet)	
Alpha	A	0	1.3	G Main seq.	23.6 (AB)
Centauri	B	1		K Main seq.	
	C	11		M Main seq.	10,400 (AC)
Barnard's	A	10	1.8	M Main seq.	1.3
Star	B	?		(Planet?)	
Wolf 359		14	2.3	M Main seq.	—
+36°2147		8	2.5	M Main seq.	0.07
		?		? (Black dwarf?)	
Sirius	A	-1	2.7	A Main seq.	19.9
	B	9		— White dwarf	
L 726-8	A	12	2.7	M Main seq.	10.9
	B	13		M Main seq.	
Ross 154		11	2.9	M Main seq.	—
Ross 248		12	3.2	M Main seq.	—
L 789-6		12	3.3	M Main seq.	—
ε Eridani		4	3.3	K Main seq.	—
Ross 128		11	3.3	M Main seq.	—
61 Cygni	A	5	3.4	K Main seq.	85 (AB)
	B	6		K Main seq.	
	C	?		? (Black dwarf?)	
ε Indi		5	3.4	K Main seq.	—

Table 8 (*Continued*)
(In order of distance)

Name	Component (If multiple)	Apparent Visual Magnitude	Distance (parsecs)	Spectral Type		Semi-Major Axis in Multiple Systems (AU)
Procyon	A	0	3.5	F	Main seq.	15.7
	B	11		—	White dwarf	
+59°1915	A	9	3.5	M	Main seq.	60
	B	10		M	Main seq.	
+43°44	A	8	3.5	M	Main seq.	156 (AB)
	B	11		M	Main seq.	
	C	?		K?	Main seq.	
−36°15693		7	3.6	M	Main seq.	—
τ Ceti		4	3.6	G	Main seq.	—
+5°1668	A	10	3.7	M	Main seq.	?
	B	?		?	?	
−39°14192		7	3.8	M	Main seq.	—
Kapteyn's Star		9	3.9	M	Main seq.	—
Kruger 60	A	10	4.0	M	Main seq.	9.5 (AB)
	B	11		M	Main seq.	
	C	?		?	(Black dwarf?)	
Ross 614	A	11	4.0	M	Main seq.	3.9
	B	14		?	Main seq. (?)	

Data adapted from: C. W. Allen (1973) Astrophysical Quantities, 3rd Ed. (London: Athlone Press)

Table 9. The 30 Brightest Stars
(In order of brightness)

Name		Apparent Visual Magnitude	Distance (parsecs)	Spectral Type		No. Known Companion Stars
Sirius	α CMa	−1.4	3	A	Main Seq.	1
Canopus	α Car	−0.7	60	F	Supergiant	0
Rigel Kent	α Cen	−0.1	1	G	Main Seq.	2
Arcturus	α Boo	−0.1	11	K	Red Giant	0
Vega	α Lyr	0.0	8	A	Main Seq.	0
Capella	α Aur	+0.1	14	G	Red Giant	1
Rigel	β Ori	0.1	250	B	Supergiant	2?
Procyon	α CMi	0.4	4	F	Subgiant	2
Achernar	α Eri	0.5	39	B	Main Seq.	0
Hadar	β Cen	0.6	120	B	Supergiant	1?
Altair	α Aql	0.8	5	A	Main Seq.	0
Betelgeuse	α Ori	0.8	200	M	Supergiant	1
Aldebaran	α Tau	0.8	21	K	Giant	2?
Acrux	α Cru	0.9	80	B	Supergiant	3?
Spica	α Vir	1.0	80	B	Main Seq.	1
Antares	α Sco	1.0	130	M	Supergiant	1
Pollux	β Gem	1.2	11	K	Red Giant	0
Fomalhout	α PsA	1.2	7	A	Main Seq.	0
Deneb	α Cyg	1.2	500	A	Supergiant	0
Mimosa	β Cru	1.3	49	B	Giant	0
Regulus	α Leo	1.4	26	B	Main Seq.	2?
Adhara	ε CMa	1.5	200	B	Supergiant	1?
Castor	α Gem	1.6	14	A	Main Seq.	5
Shaula	λ Sco	1.6	100	B	Main Seq.	1
Bellatrix	γ Ori	1.6	93	B	Giant	0
Gacrux	γ Cru	1.6	70	M	Supergiant	0
El Nath	β Tau	1.6	55	B	Giant	0
Miaplacidus	β Car	1.7	26	A	Giant	0
Alnilam	ε Ori	1.7	470	B	Supergiant	0
Al Na'ir	α Gru	1.7	21	B	Main Seq.	0

Data adapted from C. W. Allen (1973) Astrophysical Quantities, 3rd ed. (London: Athlone Press).

Table 10. Some Interesting Double Stars

Star	R.A. h. m.	Decl. ° '	Magnitudes	Position Angle, °	Distance "
Beta Tucanæ	00 29.3	−63 14	4.5, 4.5	170	27.1
Eta Cassiopeiæ	00 46.1	+57 33	3.7, 7.4	278	8.7
Beta Phœnicis	01 03.9	−46 39	4.1, 4.2	350	1.3
Zeta Phœnicis	01 06.3	−55 31	4.1, 8.4	245	6.8
Zeta Piscium	01 11.1	+07 19	4.2, 5.3	063	23.6
Alpha Ursæ Minoris	01 48.8	+89 02	2.1, 9.0	217	18.3
Gamma Arietis	01 50.8	+19 03	4.4, 4.4	000	8.4
Alpha Piscium	01 59.4	+02 31	4.3, 5.2	306	2.5
Gamma Andromedæ	02 00.8	+42 06	2.9, 5.0	061	9.7
66 Ceti	02 10.2	−02 38	5.9, 7.8	232	16.3
Gamma Ceti	02 40.7	+03 02	3.7, 6.2	293	3.0
Iota Cassiopeiæ	02 24.9	+67 11	4.2, 7.1, 8.1	215, 113	2.4, 7.4
Eta Persei	02 47.0	+55 41	4.0, 8.5	301	28.4
Epsilon Arietis	02 56.4	+21 08	6.0, 6.4	205	1.5
Theta Eridani	02 56.4	−40 30	3.4, 4.4	087	8.2
Epsilon Persei	03 54.5	+39 52	3.1, 8.3	010	9.0
Alpha Tauri	04 33.0	+16 25	0.9, 11.1	034	121
Beta Orionis	05 12.1	−08 15	0.1, 6.7	202	9.4
Eta Orionis	05 22.0	−02 26	3.8, 4.8	079	1.4
Beta Leporis	05 26.1	−20 48	3.1, 9.6	313	2.5
Delta Orionis	05 29.4	−00 20	var., 6.8	000	52.8
Alpha Leporis	05 30.5	−17 31	4.0, 9.5	156	35.5
Lambda Orionis	05 32.4	+09 54	4.0, 5.9	043	4.2
Iota Orionis	05 33.0	−05 56	3.2, 7.3	141	11.4
Theta Orionis	05 33.0	−05 27	6.0, 7.0, 7.5, 8.0		—
Sigma Orionis	05 36.2	−02 36	4.0, 10.0, 7.0, 7.5		—
Zeta Orionis	05 38.2	−01 57	1.9, 5.0	159	2.8
Theta Aurigæ	05 56.3	+37 13	2.7, 7.3	332	2.8
Gamma Volantis	07 09.2	−70 25	3.9, 5.8	299	13.7
Lambda Geminorum	07 15.2	+16 38	3.2, 10.2	033	9.9
Delta Geminorum	07 17.1	+22 05	3.2, 8.2	211	6.7
Alpha Geminorum	07 31.4	+32 00	1.9, 2.8	204	3.9
Kappa Geminorum	07 41.4	+24 31	4.0, 8.5	236	6.8
Zeta Volantis	07 42.3	−72 29	3.9, 8.9	116	16.7
Zeta Cancri	08 09.3	+17 48	5.0, 5.7	108	1.1
Delta Argûs	08 43.3	−54 31	1.9, 6.6	160	3.0
Iota Cancri	08 43.7	+28 57	4.4, 6.5	307	30.7
Epsilon Hydrae	08 44.2	+06 36	3.9, 7.8	253	3.6
Upsilon Argûs	09 45.9	−64 50	3.2, 6.0	128	5.0
Gamma Leonis	10 17.2	+20 06	2.3, 3.8	116	3.7
Xi Ursæ Majoris	11 15.6	+31 50	4.4, 4.9	292	1.9
Iota Leonis	11 21.3	+10 48	3.9, 7.1	015	0.7
Beta Hydræ	11 50.4	−33 38	4.4, 4.8	000	1.2
Alpha Crucis	12 23.7	−62 49	1.4, 1.9	119	4.7

Star	R.A. h. m.	Decl. ° '	Magnitudes	Position Angle, °	Distance "
Delta Corvi	12 27.3	−16 15	3.0, 8.5	212	24.2
Gamma Centauri	12 38.8	−48 41	3.1, 3.1	023	0.5
Gamma Virginis	12 39.1	−01 10	3.7, 3.7	317	5.5
Beta Muscæ	12 43.2	−67 49	3.9, 4.2	004	1.5
Alpha Canum Venaticic	12 53.7	+38 35	3.1, 5.7	228	19.7
Theta Virginis	13 07.4	−05 16	4.0, 8.9	343	7.2
Zeta Ursæ Majoris	13 21.9	+55 11	2.2, 4.2	150	14.5
Kappa Boötis	14 11.7	+52 01	5.0, 7.2	237	13.2
Iota Boötis	14 14.4	+51 36	4.9, 7.5	033	38.4
Alpha Centauri	14 36.6	−60 38	0.0, 1.7	310	4.0
Alpha Circini	14 38.5	−64 45	3.4, 8.8	235	15.8
Zeta Boötis	14 38.8	+13 57	4.3, 4.8	133	1.1
Epsilon Boötis	14 42.8	+27 17	3.0, 6.3	334	2.8
Delta Boötis	15 13.5	+33 30	3.2, 7.4	079	105
Eta Coronæ Borealis	15 21.1	+30 28	5.2, 5.7	280	1.0
Delta Serpentis	15 32.4	+10 42	3.0, 4.0	181	3.6
Zeta Coronæ Borealis	15 37.5	+36 48	4.0, 5.9	304	6.3
Beta Scorpionis	16 02.5	−19 40	3.0, 5.2	023	13.8
Alpha Scorpionis	16 26.4	−26 19	0.9, 6.8	275	3.0
Zeta Herculis	16 39.4	+31 41	3.0, 6.5	240	1.6
Mu Draconis	17 04.3	+54 32	5.0, 5.1	102	2.3
Alpha Herculis	17 12.4	+14 27	var., 6.1	112	4.4
Delta Herculis	17 13.0	+24 54	3.1, 7.5	208	11.0
Nu Draconis	17 31.2	+55 13	4.6, 4.6	312	62.0
Alpha Lyræ	18 35.2	+38 44	0.0, 10.5	169	56.6
Epsilon¹ Lyræ	18 42.7	+39 37	4.6, 6.3	005	2.9
Zeta Lyræ	18 43.0	+37 33	4.2, 5.5	150	43.7
Theta Serpentis	18 53.7	+04 08	4.1, 4.1	103	22.3
Eta Lyræ	19 12.0	+39 03	4.0, 8.0	083	28.2
Beta Cygni	19 28.7	+27 51	3.0, 5.3	055	34.6
Delta Cygni	19 43.5	+45 00	3.0, 7.9	263	1.9
Epsilon Draconis	19 48.3	+70 08	4.0, 7.6	009	3.3
Alpha Capricorni	20 14.9	−12 40	3.2, 4.2	291	376
Gamma Delphini	20 44.4	+15 57	4.0, 5.0	270	10.5
Beta Cephei	21 28.0	+70 20	3.4, 8.0	250	13.7
Kappa Pegasi	21 42.4	+25 25	3.9, 10.9	296	12.9
Zeta Aquarii	22 26.2	−00 17	4.4, 4.6	291	2.6
Delta Cephei	22 27.3	+58 10	var., 7.5	192	41.0

Table 11. Some Interesting Variable Stars

Star	R.A. h. m.	Decl. ° '	Magnitude Max.	Min.	Spectrum	Type	Period Days
T Ceti	00 19.2	−20 20	5.1	7.0	M	Irregular	—
R Andromedæ	00 21.4	+38 18	5.6	15	M	Long-period	410
Alpha Cassiopeiæ	00 37.6	+56 15	2.1	2.5	K	Irregular	—
Gamma Cassiopeiæ	00 49.5	+60 04	1.6	3.4	Peculiar	Pseudo-nova	—
Omicron Ceti	02 16.8	−03 12	1.6	9.6	M	Long-period	331
R Trianguli	02 34.0	+34 03	5.8	12	M	Long-period	270
Rho Persei	03 02.0	+38 39	3.3	4.1	M	Irregular	—
Beta Persei	03 04.9	+40 46	2.3	3.5	B	Eclipsing	2.87
Lambda Tauri	03 57.8	+12 20	3.3	4.2	B	Eclipsing	3.9
R Doradus	04 36.3	−62 10	5.7	6.8	M	Long-period	360
Epsilon Aurigæ	04 58.4	+43 44	3.3	4.1	F	Eclipsing	27.5 yrs.
Alpha Orionis	05 52.5	+07 24	0.1	1.3	M	Irregular	—
U Orionis	05 52.9	+20 11	5.4	12	M	Long-period	374
Eta Geminorum	06 11.9	+22 31	3.2	4.2	M	Long-period	231
T Monocrotis	06 22.5	+07 07	5.8	6.8	G	Cepheid	27.0
Zeta Geminorum	07 01.2	+20 39	3.7	4.3	G	Cepheid	10.2
R Geminorum	07 04.3	+22 48	5.9	14	G	Long-period	370
R Canis Majoris	07 17.2	−16 18	5.9	6.7	F	Eclipsing	1.14
V Puppis	07 56.7	−49 06	4.1	4.9	B	Eclipsing	1.45
R Carinae	09 31.0	−62 34	4.5	10	M	Long-period	309
l Carinae	09 43.9	−62 17	3.6	5.0	G	Cepheid	35.5
R Leonis	09 44.9	+11 40	4.9	10.5	M	Long-period	312
S Carinæ	10 07.8	−61 19	5.8	9.0	M	Long-period	149
U Hydræ	10 35.1	−13 07	4.5	6.0	N	Irregular	—
Eta Argûs	10 43.0	−59 25	−1.1	7.8	Peculiar	Pseudo-nova	—
T Ursæ Majoris	12 34.1	+59 46	5.5	13	M	Long-period	254
R Hydræ	13 26.9	−23 01	4.0	10	M	Long-period	415
S Virginis	13 30.4	−06 56	5.6	12.5	M	Long-period	372

Name	R.A.	Dec.	Mag.		Spectrum	Type	Period
T Centauri	13 38.9	−33 21	5.2	10	M	Long-period	90
Theta Apodis	14 00.5	−76 33	5.1	6.6	M	Irregular	—
R Centauri	14 13.0	−59 41	5.3	13	M	Long-period	560
W Boötis	14 41.2	+26 44	5.2	6.1	K	Irregular	—
Delta Libræ	14 58.3	−08 19	4.8	6.2	A	Eclipsing	2.05
R Coronæ Borealis	15 46.4	+28 19	5.8	12.5	Peculiar	Irregular	—
R Serpentis	15 48.4	+15 17	5.5	13.4	M	Long-period	357
T Coronæ Borealis	15 57.4	+26 04	1.9	9.5	Peculiar	Recurrent nova	—
g Herculis	16 27.0	+41 59	4.7	6.0	M	Irregular	—
S Herculis	16 49.7	+15 02	5.9	12.5	M	Long-period	300
R Scorpionis	16 53.4	−30 30	5.6	11.3	M	Long-period	279
Alpha Herculis	17 12.4	+14 27	3.1	3.9	M	Irregular	—
U Ophiuchi	17 14.0	+01 16	5.7	6.7	B	Eclipsing	1.68
u Herculis	17 15.5	+33 09	4.8	5.4	B	Eclipsing	2.05
X Sagittarii	17 44.5	−27 49	4.3	5.0	F	Cepheid	7.01
W Sagittarii	18 01.8	−29 35	4.8	5.8	F	Cepheid	7.59
V Puppis	07 56.7	−49 06	4.1	4.9	B	Eclipsing	1.45
R Carinæ	09 31.0	−62 34	4.5	10	M	Long-period	309
l Carinæ	09 43.9	−62 17	3.6	5.0	G	Cepheid	35.5
R Leonis	09 44.9	+11 40	4.9	10.5	M	Long-period	312
S Carinæ	10 07.8	−61 19	5.8	9.0	M	Long-period	149
U Hydræ	10 35.1	−13 07	4.5	6.0	N	Irregular	—
Eta Argús	10 43.0	−59 25	−1.1	7.8	Peculiar	Pseudo-nova	—
T Ursæ Majoris	12 34.1	+59 46	5.5	13	M	Long-period	254
R Hydræ	13 26.9	−23 01	4.0	10	M	Long-period	415
S Virginis	13 30.4	−06 56	5.6	12.5	M	Long-period	372
T Centauri	13 38.9	−33 21	5.2	10	M	Long-period	90
Theta Apodis	14 00.5	−76 33	5.1	6.6	M	Irregular	—
R Centauri	14 13.0	−59 41	5.3	13	M	Long-period	560
W Boötis	14 41.2	+26 44	5.2	6.1	K	Irregular	—
Delta Libræ	14 58.3	−08 19	4.8	6.2	A	Eclipsing	2.05
R Coronæ Borealis	15 46.4	+28 19	5.8	12.5	Peculiar	Irregular	—

Table 11 (*Continued*)

Star	R.A. h. m.	Decl. ° ′	Magnitude Max.	Min.	Spectrum	Type	Period Days
R Serpentis	15 48.4	+15 17	5.5	13.4	M	Long-period	357
T Coronae Borealis	15 57.4	+26 04	1.9	9.5	Peculiar	Recurrent nova	—
g Herculis	16 27.0	+41 59	4.7	6.0	M	Irregular	—
S Herculis	16 49.7	+15 02	5.9	12.5	M	Long-period	300
R Scorpionis	16 53.4	−30 30	5.6	11.3	M	Long-period	279
Alpha Herculis	17 12.4	+14 27	3.1	3.9	M	Irregular	—
U Ophiuchi	17 14.0	+01 16	5.7	6.7	B	Eclipsing	1.68
u Herculis	17 15.5	+33 09	4.8	5.4	B	Eclipsing	2.05
X Sagittarii	17 44.5	−27 49	4.3	5.0	F	Cepheid	7.01
W Sagitarii	18 01.8	−29 35	4.8	5.8	F	Cepheid	7.59

2132232

22332

Table 12. Some Interesting Clusters, Nebulae, and Galaxies Visible with Small Telescopes

Object	Constellation	R.A. h. m.	Decl. ° ′	Type
47 Tucanæ	Tucana	00 21.9	−72 22	Globular cluster
M 31	Andromeda	00 40.0	+41 00	Spiral galaxy
NGC 362	Tucana	01 00.7	−71 06	Globular cluster
M 33	Triangulum	01 31.0	+30 24	Spiral galaxy
H IV 33-34	Perseus	02 19.0	+56 54	Open clusters
M 103	Cassiopeia	01 29.8	+60 26	Open cluster
M 77	Cetus	02 40.1	−00 15	Spiral galaxy
M 45	Taurus	03 40.0	+23 30	Open cluster
M 38	Auriga	05 25.3	+35 48	Open cluster
M 1	Taurus	05 31.5	+21 59	Supernova debris
M 42	Orion	05 32.5	−05 25	Emission nebula
30 Doradûs	Dorado	05 39.1	−69 09	Emission nebula
M 37	Auriga	05 49.6	+32 33	Open cluster
M 35	Gemini	06 05.7	+24 21	Open cluster
M 41	Canis Major	06 44.9	−20 42	Open cluster
M 46	Argo Navis	07 39.5	−14 22	Open cluster
M 44	Cancer	08 37.2	+20 10	Open cluster
M 81	Ursa Major	09 51.5	+69 18	Spiral galaxy
NGC 3372	Argo Navis	10 43.0	−59 25	Emission nebula
M 97	Ursa Major	11 11.8	+55 17	Planetary nebula
NGC 3766	Centaurus	11 33.9	−61 20	Open cluster
NGC 4755	Crux Australis	12 50.7	−60 05	Open cluster
Omega Centauri	Centaurus	13 23.7	−47 03	Globular cluster
M 51	Canes Venatici	13 27.8	+47 27	Spiral galaxy
M 3	Canes Venatici	13 39.9	+28 38	Globular cluster
M 5	Serpens	15 15.9	+02 16	Globular cluster
NGC 6025	Triang. Aust.	15 59.4	−60 21	Open cluster
NGC 6067	Norma	16 09.4	−54 05	Open cluster
M 80	Scorpio	16 14.1	−22 51	Globular cluster
M 13	Hercules	16 39.9	+36 33	Globular cluster
M 19	Ophiuchus	16 59.5	−26 12	Globular cluster
M 92	Hercules	17 15.6	+43 12	Globular cluster
M 6	Scorpio	17 36.7	−32 10	Open cluster
M 23	Sagittarius	17 54.0	−19 01	Open cluster
H IV 37	Draco	17 58.6	+66 38	Planetary nebula
M 8	Sagittarius	18 00.6	−24 23	Emission nebula
NGC 6572	Ophiuchus	18 10.2	+06 50	Planetary nebula
M 17	Sagittarius	18 18.0	−16 12	Emission nebula
M 22	Sagittarius	18 23.3	−23 57	Globular cluster
M 11	Scutum	18 48.2	−06 20	Open cluster
M 57	Lyra	18 52.0	+32 58	Planetary nebula
M 27	Vulpecula	19 57.4	+22 35	Planetary nebula
H IV 1	Aquarius	21 01.4	−11 34	Planetary nebula
M 15	Pegasus	21 27.6	+11 57	Globular cluster
M 2	Aquarius	21 30.9	−01 04	Globular cluster
H IV 18	Andromeda	22 23.4	+42 12	Planetary nebula

Table 13. The Messier Catalog of Interesting Celestial Objects

Op Cl = open cluster; Glob = globular cluster; Plan = planetary nebula; Neb = diffuse or emission nebula; Gal. = galaxy (with classification).

Messier	NGC IC	Type	Con-stellation	Coordinates 1950 α (h m)	δ (° ′)	Apparent Magnitude m_v	Name, etc.
M 1	1952	Crab	Tau	05 31.5	+21 59	8.4	Crab neb
2	7089	Glob	Aqr	21 30.9	−01 03	6.3	
3	5272	Glob	CVn	13 39.9	+28 38	6.2	
4	6121	Glob	Sco	16 20.6	−26 24	6.1	
5	5904	Glob	Ser	15 16.0	+02 16	6.0	
6	6405	Op Cl	Sco	17 36.8	−32 11	5.5	
7	6475	Op Cl	Sco	17 50.7	−34 48	5	
8	6523	Neb	Sgr	18 01.6	−24 20	5.8	Lagoon neb
9	6333	Glob	Oph	17 16.2	−18 28	7.6	
10	6254	Glob	Oph	16 54.5	−04 02	6.4	
11	6705	Op Cl	Sct	18 48.4	−06 20	6.5	
12	6218	Glob	Oph	16 44.6	−01 52	6.7	
13	6205	Glob	Her	16 39.9	+36 33	5.8	
14	6402	Glob	Oph	17 35.0	−03 13	7.8	
15	7078	Glob	Peg	21 27.6	+11 57	6.3	
16	6611	Op Cl	Ser	18 16.0	−13 48	6.5	
17	6618	Neb	Sgr	18 18.0	−16 12	7	Omega neb
18	6613	Op Cl	Sgr	18 17.0	−17 09	7.2	
19	6273	Glob	Oph	16 59.5	−26 11	6.9	
20	6514	Neb	Sgr	17 58.9	−23 02	8.5	Trifid neb
21	6531	Op Cl	Sgr	18 01.8	−22 30	6.5	
22	6656	Glob	Sgr	18 33.3	−23 58	5.3	

				RA	Dec		
23	6494	Op Cl	Sgr	17 54.0	−19 01	6.5	
24	6603	Op Cl	Sgr	18 15.5	−18 27	5	
25	I4725	Op Cl	Sgr	18 28.8	−19 17	6	
26	6694	Op Cl	Sct	18 42.5	−09 27	9.1	
27	6853	Plan	Vul	19 57.4	+22 35	8.1	Dumbbell neb
28	6626	Glob	Sgr	18 21.5	−24 54	7.1	
29	6913	Op Cl	Cyg	20 22.2	+38 21	7.2	
30	7099	Glob	Cap	21 37.5	−23 25	7.7	
31	224	Gal Sb	And	00 40.0	+41 00	4.0	Andromeda neb
32	221	Gal E	And	00 40.0	+40 36	8.5	
33	598	Gal Sc	Tri	01 31.1	+30 24	6.0	
34	1039	Op Cl	Per	02 38.8	+42 34	5.7	
35	2168	Op Cl	Gem	06 05.7	+24 20	5.6	
36	1960	Op Cl	Aur	05 32.0	+34 07	6.0	
37	2099	Op Cl	Aur	05 49.0	+32 23	6.0	
38	1912	Op Cl	Aur	05 25.3	+35 48	7	
39	7092	Op Cl	Cyg	21 30.4	+48 13	5	
40	—	2 stars	UMa	12 33.0	+58 30		
41	2287	Op Cl	CMa	06 44.9	−20 42	5	
42	1976	Neb	Ori	05 32.9	−05 25	4	Orion neb
43	1982	Neb	Ori	05 33.1	−05 18	9	Orion neb
44	2632	Op Cl	Cnc	08 37.5	+19 52	3.7	Praesepe
45	—	Op Cl	Tau	03 43.9	+23 58	1.6	Pleiades
46	2437	Op Cl	Pup	07 39.6	−14 42	6	
47	2422	Op Cl	Pup	07 34.3	−14 22	5	
48	2548	Op Cl	Hya	08 11.3	−05 39	6	
49	4472	Gal E	Vir	12 27.3	+08 16	8.9	
50	2323	Op Cl	Mon	07 00.5	−08 16	6.5	
51	5194	Gal Sc	CVn	13 27.8	+47 27	8.4	Whirlpool
52	7654	Op Cl	Cas	23 22.0	+61 20	7.1	

Table 13 (*Continued*)

Op Cl = open cluster; Glob = globular cluster; Plan = planetary nebula; Neb = diffuse or emission nebula; Gal. = galaxy (with classification).

Messier	NGC IC	Type	Con- stellation	Coordinates 1950 α	δ	Apparent Magnitude m_v	Name, etc.
53	5024	Glob	Com	13 10.5	+18 26	7.7	
54	6715	Glob	Sgr	18 52.0	−30 32	7.7	
55	6809	Glob	Sgr	19 36.9	−31 03	6.1	
56	6779	Glob	Lyr	19 14.6	+30 05	8.3	
57	6720	Plan	Lyr	18 51.7	+32 58	9.0	Ring neb
58	4579	Gal SBb	Vir	12 35.1	+12 05	9.9	
59	4621	Gal E	Vir	12 39.5	+11 55	10.2	
60	4649	Gal E	Vir	12 41.1	+11 48	9.2	
61	4303	Gal Sc	Vir	12 19.4	+04 45	9.8	
62	6266	Glob	Oph	16 58.1	−30 03	7.1	
63	5055	Gal Sb	CVn	13 13.5	+42 17	8.9	
64	4826	Gal Sb	Com	12 54.3	+21 47	8.7	
65	3623	Gal Sa	Leo	11 16.3	+13 23	9.6	
66	3627	Gal Sb	Leo	11 17.6	+13 17	9.1	
67	2682	Op Cl	Cnc	08 48.3	+12 00	6.3	
68	4590	Glob	Hya	12 36.8	−26 29	8.0	
69	6637	Glob	Sgr	18 28.1	−32 23	7.8	
70	6681	Glob	Sgr	18 40.0	−32 21	8.3	
71	6838	Glob	Sge	19 51.5	+18 39	7.5	
72	6981	Glob	Aqr	20 50.7	−12 44	9.2	
73	6994	Op Cl	Aqr	20 56.4	−12 50		
74	628	Gal Sc	Psc	01 34.0	+15 32	9.6	
75	6864	Glob	Sgr	20 03.2	−22 04	8.3	

M	NGC	Type	Con	RA	Dec	Mag	Name
76	650	Plan	Per	01 38.8	+51 19	11.5	
77	1068	Gal Sb	Cet	02 40.1	−00 14	9.1	
78	2068	Neb	Ori	05 44.2	+00 02	7.4	
79	1904	Glob	Lep	05 22.2	−24 34	7.2	
80	6093	Glob	Sco	16 14.1	−22 52	7.0	
81	3031	Gal Sb	UMa	09 51.5	+69 18	8.7	
82	3034	Gal Irr	UMa	09 51.9	+69 56	7.6	
83	5236	Gal Sc	Hya	13 34.3	−29 37	9.7	
84	4374	Gal E	Vir	12 22.6	+13 10	9.5	
85	4382	Gal So	Com	12 22.8	+18 28	9.8	
86	4406	Gal E	Vir	12 23.7	+13 13	9.3	
87	4486	Gal Ep	Vir	12 28.3	+12 40	9.8	Radio gal
88	4501	Gal Ep	Com	12 29.5	+14 42	10.2	
89	4552	Gal E	Vir	12 33.1	+12 50	9.7	
90	4569	Gal Sb	Vir	12 34.3	+13 26	10.3	
91	4567	Gal S	Com	12 34.0	+11 32	6.3	
92	6341	Glob	Her	17 15.6	+43 12	6	
93	2447	Op Cl	Pup	07 42.4	−23 45	8.1	
94	4736	Gal Sb	CVn	12 48.6	+41 23	9.9	
95	3351	Gal SBb	Leo	10 41.3	+11 58	9.4	
96	3368	Gal Sa	Leo	10 44.2	+12 05	11.2	
97	3587	Plan	UMa	11 12.0	+55 18	10.4	Owl neb
98	4192	Gal Sb	Com	12 11.3	+15 11	9.9	
99	4254	Gal Sc	Com	12 16.3	+14 42	9.8	
100	4321	Gal Sc	Com	12 20.4	+16 06	8.2	
101	5457	Gal Sc	UMa	14 01.4	+54 35	10.5	Pinwheel
102	5866	Gal Sa	Dra	15 05.1	+55 57	7	
103	581	Op Cl	Cas	01 29.9	+60 27	8	
104	4594	Gal Sa	Vir	12 37.3	−11 21	9.5	Sombrero
105	3379	Gal E	Leo	10 45.2	+12 51		

Table 13 (*Continued*)

Op Cl = open cluster; Glob = globular cluster; Plan = planetary nebula; Neb = diffuse or emission nebula; Gal. = galaxy (with classification).

Messier	NGC IC	Type	Con- stellation	Coordinates 1950		Apparent Magnitude m_v	Name, etc.
				α	δ		
106	4258	Gal Sb	CVn	12 16.5	+47 35	9	
107	6171	Glob	Oph	16 29.7	−12 57	9	
108	3556	Gal Sb	UMa	11 08.7	+55 57	10.5	
109	3992	Gal SBc	UMa	11 55.0	+53 39	10.6	

Data from C. J. Allen (1973) Astrophysical Quantities, 3rd ed. (London: Athlone Press)

Table 14. Physical Properties of Selected Star Clusters

Name	Distance (parsecs)	Diameter (parsecs)	Estimated Mass (M.)	Estimated Age (years)	Comments
Open Clusters					
Ursa Major	21	7	300	2×10^8	So close that it covers large part of sky
Hyades	42	5	300	6×10^8	Naked eye object
Pleiades	127	4	350	5×10^7	Naked eye object
Praesepe	159	4	300	4×10^8	
M 67	830	4	150	4×10^9	
M 11	1,710	6	250	8×10^7	
h Persei	2,250	16	1,000	1×10^7	} "Double
χ Persei	2,400	14	900	1×10^7	} cluster"
Globular Clusters					
M 4	2,800	9	60,000	$\sim 1 \times 10^{10}$	
M 22	3,100	9	7,000,000	$\sim 1 \times 10^{10}$	
47 Tuc	4,600	5	?	$\sim 1 \times 10^{10}$	
M 13	8,200	11	300,000	$\sim 1 \times 10^{10}$	
M 5	9,200	12	60,000	$\sim 1 \times 10^{10}$	
M 3	13,000	13	210,000	$\sim 1 \times 10^{10}$	

Data adapted from W. K. Hartmann (1978) Astronomy: The Cosmic Journey, (Belmont, CA: Wadsworth).

Table 15. Physical Properties of Selected Nebulae

Name	Constellation	Approx. Distance from Earth (parsecs)	Approx. Diameter (parsecs)	Mass (M.)	Spectral Type of Associated Star
Nebulae Probably Associated with Young Objects					
Hubble's (R Mon)	Monoceros	700	10^{-5}	10^{-1}	F
Kleinmann-Low IR	Orion	500	0.1	100	?
Dark Nebulae					
Coalsack	Crux	170	8	15	none
Horsehead	Orion	350	3	0.6	B
Emission Nebulae					
Orion (central)	Orion	460	5	300	O
Eta Carinae	Carina	2,400	80	1,000	peculiar
Lagoon	Sagittarius	1,200	9	1,000	O
Trifid	Sagittarius	1,000	4	1,000	O
Reflection Nebulae					
Pleiades	Taurus	126	1.5	?	B
Cocoon	Cygnus	1,600	2	7	B
Planetary Nebulae					
Ring	Lyra	700	0.2	0.2	white dwarf?
Dumbbell	Vulpecula	220	0.3	0.2	white dwarf?
Helix	Aquarius	140	0.5	0.2	white dwarf?
Supernovae Remnants					
Crab	Taurus	2,200	3	0.1	pulsar
Veil (Loop)	Cygnus	500	22	?	?
Gum	Puppis-Vela	460	360	100,000	pulsar?

Data from: C. J. Allen (1973) Astrophysical Quantities, 3rd ed. (London: Athlone Press)
S. Maran, J. Brandt, and T. Stecher eds. (1973) The Gum Nebula and Related Problems (Washington, D.C.: NASA SP-332).
W. K. Hartman (1978) Astronomy: The Cosmic Journey, (Belmont, CA: Wadsworth).

Table 16. The Local Group of Galaxies
(Known galaxies out to a distance of 1000 kiloparsecs)

Name	Catalog Number	Distance (kpc)	Diameter (kpc)	Mass (M.)	Type[a]
Milky Way		9	30	2×10^{11}	Sb
Large Magellanic cloud		52	8	10^{10}	Ir
Small Magellanic cloud		63	5	2×10^{9}	Ir
Draco system		67	1	10^{5}	E
Ursa Minor system		67	2	10^{5}	E
Palomar 12		75	2		G
Capricorn (Zwicky)		80	10		G
Sculptor system		85	2	3×10^{6}	E
Palomar 4		120	2.5		G
Palomar 3		130	2		G
Fornax system		170	6	2×10^{7}	E
Leo I system		230	2	4×10^{6}	E4
Leo II system		230	1	10^{6}	E1
	NGC 6822	470	2	3×10^{8}	Ir
Andromeda companion	NGC 205	640	4	8×10^{9}	E5
	NGC 147	660	2	10^{9}	E
	NGC 185	660	3	10^{9}	E
Andromeda companion	M 32	660	2	3×10^{9}	E2
Andromeda galaxy	M 31	670	40^{b}	3×10^{11}	Sb
Andromeda I		700?	0.7		E
Andromeda II		700?	0.7		E
Andromeda III		700?	0.9		E
Andromeda IV		700?	1.2		E
Triangulum galaxy	M 33	730	18	1×10^{10}	Sc
	IC 1613	740	4	3×10^{9}	Ir
Maffei 1		1,000	?	2×10^{11}	S0

[a]S = Spiral (subtypes 0, a, b, c—see discussion later in chapter)
Ir = Irregular
E = Elliptical (subtypes 0 through 7)
G = Intergalactic globular cluster
[b]Diameter of Andromeda based on angular diameter 3.°1.
Data adapted from: W. K. Hartmann (1978) Astronomy, The Cosmic Journey (Belmont, CA: Wadsworth).